D1455714

Life-Cycle Analysis for New Energy Conversion and Storage Systems

MATERIALS RESEARCH SOCIETY
SYMPOSIUM PROCEEDINGS VOLUME 1041

Life-Cycle Analysis for New Energy Conversion and Storage Systems

Symposium held November 26–27, 2007, Boston, Massachusetts, U.S.A.

EDITORS:

Vasilis Fthenakis

Center for Life Cycle Analysis
Columbia University
New York, New York, U.S.A.
and
PV Environmental Research Center
Brookhaven National Laboratory
Upton, New York, U.S.A.

Anne Dillon

National Renewable Energy Laboratory
Golden, Colorado, U.S.A.

Nora Savage

National Center for Environmental Research
U.S. Environmental Protection Agency
Washington, D.C., U.S.A.

Materials Research Society
Warrendale, Pennsylvania

Published by:

Materials Research Society
506 Keystone Drive
Warrendale, PA 15086
Telephone (724) 779-3003
Fax (724) 779-8313
Web site: http://www.mrs.org/

Manufactured in the United States of America

CONTENTS

Preface ix

Acknowledgments xi

Materials Research Society Symposium Proceedings.....................................xii

PV ENERGY CONVERSION

* Reduction of Environmental Impacts in Crystalline Silicon
Photovoltaic Technology: An Analysis of Driving Forces
and Opportunities...3
    Erik Alsema and Mariska de Wild-Scholten

* Key Projections on Future PV Performance, Market
Penetration and Costs, with Special Reference to CdTe
and Other Thin Film Technologies 13
    Marco Raugei and Paolo Frankl

Environmental Implications of Nanostructured
Photovoltaics: A Comparative Life-Cycle Analysis
Framework...25
    V. Fthenakis, S. Gualtero, R. van der Meulen,
    and H.C. Kim

Life-Cycle Assessment of Photovoltaics: Update of the
ecoinvent Database 33
    Niels Jungbluth, Roberto Dones, and Rolf Frischknecht

* IR Based Photovoltaic Array Performance Assessment 43
    A. Moropoulou, J.A. Palyvos, M. Karoglou, and
    V. Panagopoulos

NANOMATERIALS AND HYDROGEN STORAGE

* Nano-Structured Materials to Address Challenges of the
Hydrogen Initiative...51
    Vincent Berube and Mildred Dresselhaus

*Invited Paper

* **High-Surface-Area Biocarbons for Reversible On-Board Storage of Natural Gas and Hydrogen**...63
    Peter Pfeifer, Jacob W. Burress, Mikael B. Wood,
    Cintia M. Lapilli, Sarah A. Barker, Jeffrey S. Pobst,
    Raina J. Cepel, Carlos Wexler, Parag S. Shah,
    Michael J. Gordon, Galen J. Suppes, S. Philip Buckley,
    Darren J. Radke, Jan Ilavsky, Anne C. Dillon,
    Philip A. Parilla, Michael Benham, and Michael W. Roth

* **Hydrogen Adsorption in MOF-74 Studied by Inelastic Neutron Scattering**..75
    Yun Liu, Craig M. Brown, Dan A. Neumann,
    Houria Kabbour, and Channing C. Ahn

* **Metal Hydrides for Hydrogen Storage**...85
    Jason Graetz, James J. Reilly, and James Wegrzyn

* **Science and Prospects of Using Nanoporous Materials for Energy Absorption**...95
    Xi Chen and Yu Qiao

* **Novel Organometallic Fullerene Complexes for Vehicular Hydrogen Storage**..107
    Erin Whitney, Anne C. Dillon, Calvin Curtis,
    Chaiwat Engtrakul, Kevin O'Neill, Mark Davis,
    Lin Simpson, Kim Jones, Yufeng Zhao,
    Yong-Hyun Kim, Shengbai Zhang, and Philip Parilla

NOVEL ENERGY STORAGE
TECHNOLOGIES

A Novel High Capacity, Environmental Benign Energy Storage System: Super-Iron Boride Battery 119
    Xingwen Yu and Stuart Licht

Mechanical Effect on Oxygen Mobility in Yttria Stabilized Zirconia...125
    Wakako Araki and Tadaharu Adachi

Effect of Water Vapor and SOx in Air on the Cathodes of Solid Oxide Fuel Cells...131
    Seon Hye Kim, Toshihiro Ohshima, Yusuke Shiratori,
    Kohei Itoh, and Kazunari Sasaki

*Invited Paper

**Compaction and Cold Crucible Induction Melting of Fine
Poly Silicon Powders for Economical Production of
Polycrystalline Silicon Ingot** 139
    Daesuk Kim, Jesik Shin, Byungmoon Moon, and
    Kiyoung Kim

ENERGY LCA METHODOLOGY

* **Life-Cycle Assessment of Future Fossil Technologies
with and without Carbon Capture and Storage**..147
    Roberto Dones, Christian Bauer, Thomas Heck,
    Oliver Mayer-Spohn, and Markus Blesl

**Integration of Land Use Aspects Into Life-Cycle
Assessment at the Example of Biofuels** 159
    Michael Held and Ulrike Bos

**The Fuel Cycles of Electricity Generation: A Comparison
of Land Use**...165
    Hyung Chul Kim and Vasilis Fthenakis

* **Wise Energy Investment Decisions—Not Just [kJ out/kJ in]** 173
    Lise Laurin

* **Standing the Test of Time: Signals and Noise From
Environmental Assessments of Energy Technologies**...183
    Björn A. Sandén

Author Index...191

Subject Index..193

*Invited Paper

PREFACE

Symposium R, "Life-Cycle Analysis (LCA) for New Energy Conversion and Storage Systems," held November 26–27 at the 2007 MRS Fall Meeting in Boston, Massachusetts, was the second symposium on life-cycle analysis organized under the auspices of the Materials Research Society. It commendably extended the inaugural symposium that took place during the 2005 MRS Fall Meeting. The sessions encompassed photovoltaics, nanomaterials, batteries and hybrid cars, hydrogen storage, and issues of LCA methodologies.

Life-cycle analyses, vital in detailing the range of a technology's detrimental effects upon the environment and human health, afford us solid information for judiciously selecting the least damaging ones.

The first chapter of this book presents new information on the life cycle of photovoltaics. Since this technology is rapidly evolving, periodic updates are essential for ensuring well-balanced comparisons with other technologies. Furthermore, this chapter presents a framework for assessing life cycle implications of nanostructured photovoltaics from process-based comparative analyses of the life stages of nano- and bulk- semiconductor materials.

The second section encompasses papers on nanomaterial-based technologies for energy storage that range from authoritative, comprehensive overviews of the use of nano-structured materials in hydrogen storage, to environmental issues related to such storage, and descriptions of new nano-scale systems for storing energy.

The papers in the third chapter of this volume describe some interesting new energy-storage technologies; propitiously, they provide useful technical information for life-cycle analysts.

Among the studies discussed in the fourth chapter are LCA methodologies applicable to energy systems analysis; they detail ways of quantifying land use, and suggest applicable metrics. Furthermore, they highlight the uncertainties in the LCA results and point out what is important in comparing energy technologies with LCA tools.

These papers collectively bring us a major step forward along the challenging path of generating well-balanced, accurate assessments of the environmental impacts of new energy-conversion and storage technologies. We anticipate periodically updating this latest information as emerging technologies progress to more advanced stages of commercialization.

Vasilis Fthenakis
Anne Dillon
Nora Savage

March 2008

ACKNOWLEDGMENTS

We thank the invited and contributed speakers for their efforts to prepare excellent presentations that stimulated interesting conversations during the two days of the symposium.

We also extend our gratitude to the MRS organizing committee and to the following sponsor for covering part of the travel expenses for the invited speakers of the session on PV Energy Conversion:

First Solar Inc.

MATERIALS RESEARCH SOCIETY SYMPOSIUM PROCEEDINGS

Volume 1024E —Combinatorial Methods for High-Throughput Materials Science, D.S. Ginley, M.J. Fasolka, A. Ludwig, M. Lippmaa, 2008, ISBN 978-1-60511-000-4

Volume 1025E —Nanoscale Phenomena in Functional Materials by Scanning Probe Microscopy, L. Degertekin, 2008, ISBN 978-1-60511-001-1

Volume 1026E —Quantitative Electron Microscopy for Materials Science, E. Snoeck, R. Dunin-Borkowski, J. Verbeeck, U. Dahmen, 2008, ISBN 978-1-60511-002-8

Volume 1027E —Materials in Transition—Insights from Synchrotron and Neutron Sources, C. Thompson, H.A. Dürr, M.F. Toney, D.Y. Noh, 2008, ISBN 978-1-60511-003-5

Volume 1029E —Interfaces in Organic and Molecular Electronics III, K.L. Kavanagh, 2008, ISBN 978-1-60511-005-9

Volume 1030E —Large-Area Processing and Patterning for Active Optical and Electronic Devices, V. Bulović, S. Coe-Sullivan, I.J. Kymissis, J. Rogers, M. Shtein, T. Someya, 2008, ISBN 978-1-60511-006-6

Volume 1031E —Nanostructured Solar Cells, A. Luque, A. Marti, 2008, ISBN 978-1-60511-007-3

Volume 1032E —Nanoscale Magnetic Materials and Applications, J-P. Wang, 2008, ISBN 978-1-60511-008-0

Volume 1033E —Spin-Injection and Spin-Transfer Devices, R. Allenspach, C.H. Back, B. Heinrich, 2008, ISBN 978-1-60511-009-7

Volume 1034E —Ferroelectrics, Multiferroics, and Magnetoelectrics, J.F. Scott, V. Gopalan, M. Okuyama, M. Bibes, 2008, ISBN 978-1-60511-010-3

Volume 1035E —Zinc Oxide and Related Materials—2007, D.P. Norton, C. Jagadish, I. Buyanova, G-C. Yi, 2008, ISBN 978-1-60511-011-0

Volume 1036E —Materials and Hyperintegration Challenges in Next-Generation Interconnect Technology, R. Geer, J.D. Meindl, R. Baskaran, P.M. Ajayan, E. Zschech, 2008, ISBN 978-1-60511-012-7

Volume 1037E —Materials, Integration, and Technology for Monolithic Instruments II, D. LaVan, 2008, ISBN 978-1-60511-013-4

Volume 1038— Nuclear Radiation Detection Materials, D.L. Perry, A. Burger, L. Franks, M. Schieber, 2008, ISBN 978-1-55899-985-5

Volume 1039— Diamond Electronics—Fundamentals to Applications II, R.B. Jackman, C. Nebel, R.J. Nemanich, M. Nesladek, 2008, ISBN 978-1-55899-986-2

Volume 1040E —Nitrides and Related Bulk Materials, R. Kniep, F.J. DiSalvo, R. Riedel, Z. Fisk, Y. Sugahara, 2008, ISBN 978-1-60511-014-1

Volume 1041E —Life-Cycle Analysis for New Energy Conversion and Storage Systems, V.M. Fthenakis, A.C. Dillon, N. Savage, 2008, ISBN 978-1-60511-015-8

Volume 1042E —Materials and Technology for Hydrogen Storage, G-A. Nazri, C. Ping, A. Rougier, A. Hosseinmardi, 2008, ISBN 978-1-60511-016-5

Volume 1043E —Materials Innovations for Next-Generation Nuclear Energy, R. Devanathan, R.W. Grimes, K. Yasuda, B.P. Uberuaga, C. Meis, 2008, ISBN 978-1-60511-017-2

Volume 1044— Thermoelectric Power Generation, T.P. Hogan, J. Yang, R. Funahashi, T. Tritt, 2008, ISBN 978-1-55899-987-9

Volume 1045E —Materials Science of Water Purification—2007, J. Georgiadis, R.T. Cygan, M.M. Fidalgo de Cortalezzi, T.M. Mayer, 2008, ISBN 978-1-60511-018-9

MATERIALS RESEARCH SOCIETY SYMPOSIUM PROCEEDINGS

Volume 1046E —Forum on Materials Science and Engineering Education for 2020, L.M. Bartolo, K.C. Chen,
    M. Grant Norton, G.M. Zenner, 2008, ISBN 978-1-60511-019-6
Volume 1047 — Materials Issues in Art and Archaeology VIII, P. Vandiver, F. Casadio, B. McCarthy, R.H. Tykot,
    J.L. Ruvalcaba Sil, 2008, ISBN 978-1-55899-988-6
Volume 1048E—Bulk Metallic Glasses—2007, J. Schroers, R. Busch, N. Nishiyama, M. Li, 2008,
    ISBN 978-1-60511-020-2
Volume 1049 — Fundamentals of Nanoindentation and Nanotribology IV, E. Le Bourhis, D.J. Morris, M.L. Oyen,
    R. Schwaiger, T. Staedler, 2008, ISBN 978-1-55899-989-3
Volume 1050E —Magnetic Shape Memory Alloys, E. Quandt, L. Schultz, M. Wuttig, T. Kakeshita, 2008,
    ISBN 978-1-60511-021-9
Volume 1051E —Materials for New Security and Defense Applications, J.L. Lenhart, Y.A. Elabd, M. VanLandingham,
    N. Godfrey, 2008, ISBN 978-1-60511-022-6
Volume 1052 — Microelectromechanical Systems—Materials and Devices, D. LaVan, M.G. da Silva, S.M. Spearing,
    S. Vengallatore, 2008, ISBN 978-1-55899-990-9
Volume 1053E —Phonon Engineering—Theory and Applications, S.L. Shinde, Y.J. Ding, J. Khurgin,
    G.P. Srivastava, 2008, ISBN 978-1-60511-023-3
Volume 1054E —Synthesis and Surface Engineering of Three-Dimensional Nanostructures, O. Hayden,
    K. Nielsch, N. Kovtyukhova, F. Caruso, T. Veres, 2008, ISBN 978-1-60511-024-0
Volume 1055E —Excitons and Plasmon Resonances in Nanostructures, A.O. Govorov, Z.M. Wang,
    A.L. Rogach, H. Ruda, M. Brongersma, 2008, ISBN 978-1-60511-025-7
Volume 1056E —Nanophase and Nanocomposite Materials V, S. Komarneni, K. Kaneko, J.C. Parker, P. O'Brien,
    2008, ISBN 978-1-60511-026-4
Volume 1057E —Nanotubes and Related Nanostructures, Y.K. Yap, 2008, ISBN 978-1-60511-027-1
Volume 1058E —Nanowires—Novel Assembly Concepts and Device Integration, T.S. Mayer, 2008,
    ISBN 978-1-60511-028-8
Volume 1059E —Nanoscale Pattern Formation, W.J. MoberlyChan, 2008, ISBN 978-1-60511-029-5
Volume 1060E —Bioinspired Polymer Gels and Networks, F. Horkay, N.A. Langrana, A.J. Ryan, J.D. Londono,
    2008, ISBN 978-1-60511-030-1
Volume 1061E —Biomolecular and Biologically Inspired Interfaces and Assemblies, J.B.-H. Tok, 2008,
    ISBN 978-1-60511-031-8
Volume 1062E —Protein and Peptide Engineering for Therapeutic and Functional Materials, M. Yu, S-W. Lee,
    D. Woolfson, I. Yamashita, B. Simmons, 2008, ISBN 978-1-60511-032-5
Volume 1063E —Solids at the Biological Interface, V.L. Ferguson, J.X-J. Zhang, C. Stoldt, C.P. Frick, 2008,
    ISBN 978-1-60511-033-2
Volume 1064E —Quantum-Dot and Nanoparticle Bioconjugates—Tools for Sensing and Biomedical Imaging,
    J. Cheon, H. Mattoussi, C.M. Niemeyer, G. Strouse, 2008, ISBN 978-1-60511-034-9
Volume 1065E —Electroactive and Conductive Polymers and Carbon Nanotubes for Biomedical Applications,
    X.T. Cui, D. Hoffman-Kim, S. Luebben, C.E. Schmidt, 2008, ISBN 978-1-60511-035-6

Prior Materials Research Society Symposium Proceedings available by contacting Materials Research Society

PV Energy Conversion

Mater. Res. Soc. Symp. Proc. Vol. 1041 © 2008 Materials Research Society   1041-R01-01

Reduction of Environmental Impacts in Crystalline Silicon Photovoltaic Technology: An Analysis of Driving Forces and Opportunities

Erik Alsema[1], and Mariska de Wild-Scholten[2]

[1]Copernicus Institute, Utrecht University, Heidelberglaan 2, Utrecht, Netherlands
[2]Unit Solar Energy, Energy Research Centre of the Netherlands (ECN), Westerduinweg 3, Petten, Netherlands

ABSTRACT

We give an overview of historical developments with respect to the price and the Energy Pay-Back Time of crystalline silicon photovoltaic modules. We investigate the drivers behind both developments and observe that there is a large overlap between them. Reduction of silicon consumption, improved cell efficiency and the production technology for solar grade silicon have been identified as major drivers for both cost and impact reductions in the past. Also we look into future prospects for reduction of environmental impacts. It is estimated that developments underway to reduce costs will also result in a reduction of the Energy Pay-Back Time of a PV installation (in South-Europe) from 1.5-2.0 year presently to well below 1 year.

INTRODUCTION

In this paper we will look at historical trends in the prices and in Energy Pay Back Time (EPBT) of crystalline silicon PV modules since 1975. The prices are regarded as a proxy for production costs while the EPBT can be considered as good indicator for the environmental imapcts of PV technology. We will identify the drivers that have contributed to the decrease of prices on the one hand and EPBT on the other hand, and investigate the overlap between these drivers. After this historical perspective we will look into future prospects with respect to new c-Si cell concepts and future module production technology and estimate possible reductions in the medium term.

ANALYSIS OF HISTORICAL TRENDS

Figure 1 shows the development of module prices and module production volumes since 1975 [1-3]. We have added to this graph the Energy Pay-Back Time that can be determined from a number energy analysis or Life Cycle Assessment studies that have been published over the same period [4-8]. The Energy Pay-Back Times have all been calculated for a hypothetical roof-top system in South-Europe with an array-plane irradiation of 1700 kWh/m2 /yr and a Performance Ratio of 0.75. Energy requirements for Balance-of-System (BOS) components have been assumed the same for all historical data sets, namely 70 MJ (primary energy) per m^2 module area for supports and cables and 1300 MJ/kWp for the inverter. In this way the EPBT values will reflect the developments in energy requirements for the modules only[1]. For the data set from Hunt (1976) which did not include energy inputs for the process steps of module assembly and framing we assumed for these steps the same values as given by Hagedorn (1992). Except for the data by Hunt all values are valid for modules based on multicrystalline silicon.

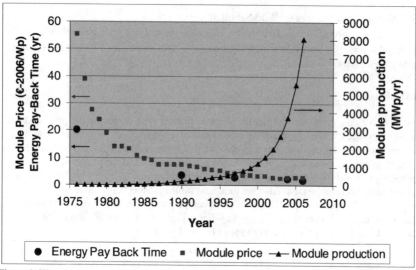

Figure 1: Historical development of module prices (squares, left y-axis, €-2006/Wp), production volumes (triangles, right y-axis; MWp/yr)) and Energy Pay-Back Time (circles, left y-axis, yr). Sources: refs. [1-8]

From the figure we can observe that since 1975 the module price has decreased by a factor 20, while production volume has grown by 25,000x and the Energy Pay-Back Time has gone down with a factor 10.
As major drivers for the cost reduction we can identify:
- Silicon consumption
- Cell efficiency
- Silicon cost
- Production scale

The next question is which drivers have been behind the reduction in EPBT? For this we need to consider more closely the technology characteristics and results of the individual EPBT studies, as given in figure 2.
One of the first thing that springs to the eye is the large energy input for silicon feedstock in the analysis by Hunt (1976), which cannot be explained only by the

---

[1] Because of these standardized BOS and system performance assumptions and some other corrections EPBT values may differ from those given by the original study. Also note that we placed EPBT values at a time value 1-2 year prior to year of publication so as to reflect better the actual year for which the data are valid.

relatively high silicon consumption of 32 g/Wp. When we also look at the energy requirements for silicon feedstock in the different studies (table I) we see that this was quite high in the Hunt study and that silicon GER values[2] remained more or less constant after that. We should add that this value has always been one of the most difficult to estimate because of the complexity of the process, allocation problems and the high confidentiality of process data (see also [Alsema, 1998]).

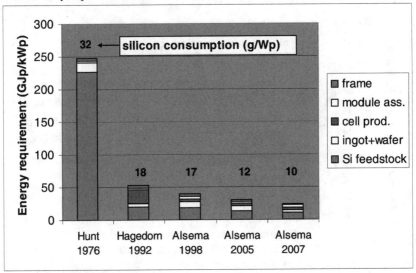

Figure 2: Break-down of energy requirements for multicrystalline silicon modules (in GJ-primary per kWp) in different energy analysis studies, together with their year of publication and the amounts of silicon feedstock required per Wp.

Table I: Gross Energy Requirement of solar grade silicon used in each EPBT study

	Gross Energy Requirements for solar-grade silicon[1] (MJ/kg)
Hunt, 1976	6700
Hagedorn, 1992	1100[2]
Alsema, 1998	1120
Alsema, 2005	1070
Alsema, 2007	1070

[1] Electronic grade silicon for the Hunt and Hagedorn studies, solar grade did not exist yet at that time.
[2] Not specified by Hagedorn, estimated by later analysists (see e.g. [9]).

Looking at the results of the period 1992-2007 in more detail (see figure 3) we observe that the feedstock part of the energy input has decreased by almost a factor 2, solely due to the decreased silicon consumption. Furthermore the cell processing part has

---

[2] GER= Gross Energy Requirement, the cumulative input of primary energy calculated over the whole production chain of a material.

decreased, probably due to increased cell efficiency (i.e. less cell area per Wp) and perhaps due to larger production scales. Between 2005 and 2007 we see a reduced energy input for the ingot and wafering process, which was mostly due to improved wafer yields and better energy efficiency in the casting process. Finally the framing part has decreased, which can mostly be attributed to increased module areas.

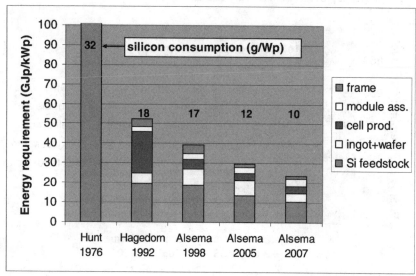

Figure 3: Same as figure 2, but with focus on results 1992-2007

What we learn from this is that the following drivers have been important in the reduction of Energy-Pay-Back Time:
- Silicon consumption
- (Energy input in) silicon feedstock production
- Cell efficiency
- Energy efficiency in casting and other process steps
- (Production scale?)

Comparing the latter set of drivers with those for the price reduction we see that at least three drivers namely silicon consumption, silicon feedstock production and cell efficiency return in both lists. Of course silicon feedstock costs and energy input in silicon feedstock production is not quite the same, but energy is certainly a considerable cost factor in silicon production (other than in other module production steps).

This overlap in the drivers for cost and energy reduction implies that efforts to reduce costs per Wp – which is the most important R&D objective – will in many cases also have beneficial effect on energy pay-back. On the other hand we can observe that production scale increase generally has an uncertain effect on energy efficiency, except in casting (lower thermal losses). At the same time it is clear improvements in energy efficiency will have small impacts on production cost in wafering, cell processing and module assembly.

ENERGY PAY-BACK TIME FOR PRESENT SITUATION

Based on our most recent Life Cycle Inventory for the three crystalline silicon technologies, as collected within the Crystal Clear project [10], we can calculate the Energy Pay-Back time for the current technology status (see fig. 4). We see that for a roof-top system in South-Europe the EPBT is respectively 1.5, 1.7 and 2.0 years, depending whether we choose ribbon, multi- and monocrystalline silicon modules.

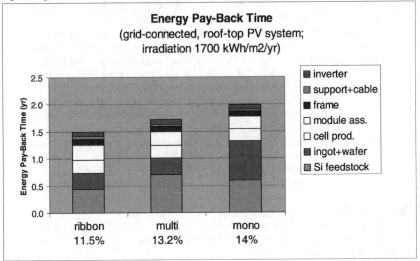

Figure 4: Energy Pay-Back Time of crystalline silicon PV systems in 2006 (rooftop system in S.-Europe, irrad. 1700 kWh/m2/yr, PR=0.75).

Of course we can also investigate which opportunities there are for a further reduction of the energy requirement for modules.

PROSPECTS FOR FUTURE REDUCTIONS OF ENERGY REQUIREMENTS

For our analysis of future prospect we will focus on improvements that relate to the four drivers of energy reduction identified above:
- Reduction of silicon consumption
- Higher cell efficiency
- New Si feedstock processes with lower energy requirements
- Reduced energy consumption for ingot growing.

New silicon feedstock processes

We have seen that the process energy for production of poly-Si is responsible for more than 30% of the total primary energy input for a multi-Si module. On average about 110 kWh of electricity and 185 MJ of heat is used to produce 1 kg of poly-Si with the improved-Siemens process that is most common at this moment. Because of this high energy consumption energy costs are a significant cost driver, so that new processes, especially for solar grade silicon, are likely to have lower energy consumption.

The process that employs Fluidized Bed Reactors to replace Siemens reactors is reported to have a much lower *electricity consumption*. Reliable quotes for this technology are hard to come by, but it seems that a reduction to 30 kWh/kg Si is possible with FBR. Heat requirements, however, remain more or less the same. The Cumulative Energy Demand of sog-silicon, produced by an improved-Siemens process is estimated at 1070 MJ/kg, while for FBR-silicon we estimate it at about 500 MJ/kg.

For direct metallurgic processes that produce solar grade silicon directly from silica, also heat requirements may be reduced because the step of gas phase distillation is omitted. A published energy estimation for this process is 25 kWh/kg ($\approx$300 MJ/kg) [11].

Figure 5 gives an indication of the effects that adoption of the FBR process instead of the Siemens process would have on the Energy Pay-Back Time of a PV system based on multicrystalline silicon modules. From the figure it is clear that new feedstock processes can give a dramatic improvement in the environmental profile of PV systems. On top of this other improvements are feasible, as is discussed below.

Barriers for the introduction of new feedstock process are the technological complexity, incomplete understanding of the allowable impurity levels and the high capital requirements for commercial scale plants. However, due to the present silicon scarcity several plants based on new process technology are now under construction.

Reduction of silicon consumption

The effect of reduced silicon consumption (in g per Wp) has been depicted in figure 5. Observe that silicon consumption has decreased significantly over the past 2 years, driven by the silicon shortage.

Obvious ways to reduce silicon consumption are:
- improved crystallization with lower loss
- thinner wafers
- lower kerf loss
- reduce wafer breakage
- recycling of silicon waste from
  o ingot cut-offs
  o broken wafers
  o kerf loss
- casting or pulling wafers directly from liquid Si (ribbon technologies)
- increased cell efficiency

Especially the recycling of ingot cut-offs and broken wafers is relatively easy. After some cleaning this material may be used as silicon feedstock in the ingot casting process.

All of the silicon reduction approaches, except kerf loss recycling, are already followed within the PV industry and most are also part of the CrystalClear activities.

Based on these activities a silicon consumption of 4-6 g/Wp seems to be well in reach within a few years [12, 13].

Kerf loss (the material cut away when sawing a wafer) is much less easy to recycle, as it is found in a mixed waste stream together with SiC particles and glycol cutting fluid. For this reason the recycling of silicon kerf loss is – to our knowledge – not done anywhere on a commercial scale, but research on it has been conducted within the FP5 project RE-Si-CLE. If such a process becomes available and does not require too much energy it could substantially reduce silicon consumption by perhaps 30-40%.

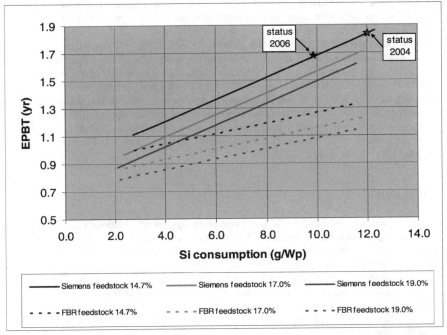

Figure 5: The Energy Pay-Back Time as a function of silicon consumption, for different combinations of a silicon feedstock process (Siemens, FBR) and multi-Si cell efficiency (resp. 14.7%, 17%, 19%). It is assumed that all other material and energy consumption for the module does not change, except that it is directly proportional to the module area. Module type: multi-Si module, frameless. System: roof-top system installed in S-Europe (1700 kWh/m2/y) with PR =0.75.

Ribbon technologies for producing wafer directly from liquid silicon are in commercial operation and require 7-8 g silicon per dm^2, but cell efficiencies are still lower than for conventional wafers at 12.5-14%, so Si consumption per Wp is 5-6 g. For this reason ribbon-Si modules currently have the lowest energy pay-back time (1.5 yr) among all silicon technologies[3].

---

[3] Note that EPBT values for ribbon technology cannot be derived from Figure 5 because this figure assumes conventional ingot and wafering processes. See [4] for EPBT values of ribbon and other cell technologies.

Barriers for (further) reduction of silicon consumption are manifold: silicon quality issues (Si recycling), sawing, cell and wafer handling (thinner wafers), cutting wire strength (kerf loss).

Increased energy-efficiency in ingot growing

From figure 4 we have seen that ingot and wafering represent a considerable part of the energy input for a module, especially for mono-Si material. At the same time we have observed that considerable differences in electricity consumption exist which mainly arise in the process of ingot growing. From the background data we also observe a tendency that newer installations have lower electricity consumption. This would imply that there is considerable scope for improvement of the energy efficiency in ingot growing.

One aspect of increased efficiency in newer facilities is probably the larger batch size, which naturally reduces energy losses from the containers of molten silicon.

When looking at the process sequence of the crystal growing process, with its cycle of melting silicon and then slowly cooling it down again, it seems sensible to investigate the possibility of heat recuperation. For example one could think of using the waste heat from the ingot that is cooled down to preheat the next batch of silicon.

Barriers for improved energy efficiency in ingot growing are probably: a lack of urgency (cost advantages unclear), a focus on material quality and long lifetimes of crystal growing equipment.

OUTLOOK

If we combine a number of improvement options which are already available or will become feasible within the next 3-5 years, we can analyse the total overall improvement that is possible. For this we focus on multicrystalline silicon technology and we assume the use of Fluidized Bed Reactor technology for silicon feedstock material, best available technology for ingot casting, 150 um wafer thickness, 17% module efficiency and no F-gas emissions. As an extra case we assume that PV operations, from ingot casting to module, will be run on "green" electricity supply, namely wind power. (The FBR feedstock process was in both cases assumed to run on hydropower[4].)

Figure 6 shows the resulting Energy Pay Back Time (EPBT) and the life-cycle greenhouse gas (GHG) emission for a roof-top PV system in South Europe (irradiation 1700 kWh/m2/yr, PR=0.75, PV system lifetime = 30 year). No improvements in BOS or in PR have been assumed.

---

[4] The present production of PV modules requires about 80 kton of silicon per year, and thus 3-10 TWh of electricity, which is only a few promille of the global hydropower generation. Nevertheless, when PV production grows by orders of magnitude other power sources will need to be tapped. We may expect, however, that by that time much more low-carbon sources like wind, solar and possibly carbon capture will have penetrated the electricity supply system.

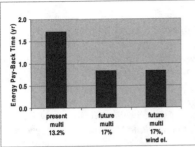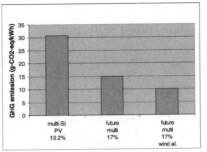

Figure 6: Potential improvements in energy pay-back time (in years, left) and greenhouse gas emissions (in g/kWh, right) for a multicrystalline silicon roof-top PV system in South Europe (irradiation 1700 kWh/m2/yr).

We can see that the EPBT can be reduced by 50%, to well below 1 year, while the case of wind electricity obviously makes no further difference for the EPBT[5]. With respect to greenhouse gas emissions the present emission of 30 g/kWh can be reduced to about 15 g/kWh, and with the additional switch to green electricity supply even to 10 g/kWh. At this latter value the GHG emission of c-Si PV technology gets in the same range as wind energy and other low-carbon energy options [14].

CONCLUSIONS

We have seen that over the last 30 years both the prices and the environmental impacts of crystalline silicon modules have been reduced substantially. It appears that the technological drivers behind both developments show a great overlap. Reduction of silicon consumption, improved cell efficiency and the production technology for solar grade silicon have been identified as major drivers for both cost and impact reduction.

Also we have reviewed a number of options to achieve a further reduction of greenhouse gas emissions in crystalline silicon module production. Altogether we have shown that there are good possibilities to reduce the Energy Pay-Back Time of a multicrystalline silicon PV system from today's 1.7 years to less than 1 year (roof-top system in South Europe). Life-cycle greenhouse gas emissions for such a system can be reduced from 30 g/kWh to 15 g/kWh or less.

ACKNOWLEDGMENTS

This study was partly conducted within the framework of the Integrated Project CrystalClear, a research and development project on advanced industrial crystalline silicon PV technology. This project is supported by European Commission under contract number SES6-CT_2003-502583.

For more information on the CrystalClear project: http://www.ipcrystalclear.info/

---

[5] The EPBT remains unchanged if we consider a wind-only electricity supply system because the electricity avoided by PV generation comes from the same wind-only system.

REFERENCES

1. Strategies Unlimited, *Photovoltaic Five-Year Market Forecast 2002-2007*, Strategies Unlimited Mountain View, California, USA, 2003.
2. Swanson, RM, *A Vision for Crystalline Silicon Photovoltaics*, Progress In Photovoltaics: Research and Applications, 2006, **14**: p. 443-453.
3. Hirschman, WP, G Hering, and M Schmela, *Market survey on global solar cell and module production in 2006*, in *Photon International*, 2007, p. 136-166.
4. Hunt, LP. *Total energy use in the production of silicon solar cells from raw materials to finished product*, in *12th IEEE Photovoltaic Specialists Conference*, Baton Rouge, LA, USA, 1976, p. 347-252.
5. Hagedorn, G. and E. Hellriegel, *Umwelrelevante Masseneinträge bei der Herstellung verschiedener Solarzellentypen - Endbericht - Teil I: Konventionelle Verfahren*, Forschungstelle für Energiewirtschaft, München, Germany, 1992.
6. Alsema, E.A., P. Frankl, and K. Kato. *Energy Pay-back Time of Photovoltaic Energy Systems: Present Status and Prospects*, in *2nd World Conference on Photovoltaic Solar Energy Conversion*, Vienna, 6-10 July, 1998
7. Alsema, E. and M.J. Wild-Scholten. *Environmental Impacts of Crystalline Silicon Photovoltaic Module Production*, in *Materials Research Society Fall 2005 Meeting*, Warrendale, USA, 2005, Materials Research Society
8. Alsema, E.A. and M.J. de Wild-Scholten. *Reduction of the Environmental Impacts in Crystalline Silicon Module Manufacturing*, in *22nd European Photovoltaic Solar Energy Conference*, Milano, 2007, WIP-Renewable Energies, Munich, Germany
9. Jungbluth, N, *Photovoltaics*, in *Sachbilanzen von Energiesystemen: Grundlagen für den ökologischen Vergleich von Energiesystemen und den Einbezug von Energiesystemen in Ökobilanzen für die Schweiz, Final report ecoinvent v2.0 No. 6*, R. Dones, Editor, 2007, Swiss Centre for Life Cycle Inventories: Duebendorf, CH.
10. Wild-Scholten, M.J. de and E.A. Alsema, *Environmental Life Cycle Inventory of Crystalline Silicon Photovoltaic Module Production, version 2, status 2005/2006*, 2007, ECN, Petten, p. Excel file, http://www.ecn.nl/docs/library/report/2007/e07026-LCIdata-cSiPV-pubv2_0.xls.
11. Friestad, K. in *19th European Photovoltaic Solar Energy Conference*, Paris, 2004
12. Aulich, HA and F-W Schulze. *Silicon for the PV industry: demand, supply and growth prospects*, in *21st European Photovoltaic Solar Energy Conference*, Dresden, 2006, p. 549-553.
13. Sinke, WC *Crystal Clear roadmap, v3, draft*, CrystalClear Consortium, 2006.
14. Alsema, E.A., M.J. de Wild-Scholten, and V Fthenakis. *Environmental Impacts of PV Electricity Generation - A Critical Comparison of Energy Supply Options*, in *21st European Photovoltaic Solar Energy Conference*, Dresden, 2006, WIP-Renewable Energies, Munich, Germany, p. 3201-3207.

Mater. Res. Soc. Symp. Proc. Vol. 1041 © 2008 Materials Research Society

Key Projections on Future PV Performance, Market Penetration and Costs, with Special Reference to CdTe and Other Thin Film Technologies

Marco Raugei[1], and Paolo Frankl[2]
[1]Ambiente Italia, Rome, Italy
[2]International Energy Agency, Paris, France

ABSTRACT

The authors have drafted three alternative scenarios for the technological improvement and market penetration of photovoltaics in the next four decades, based on the preliminary results of the EU FP6 Integrated Project NEEDS, Research Stream 1a. The long-term diffusion of PV is foreseen to depend on the achievable module efficiencies and on the maturity of the different technologies in terms of their manufacturing costs, energy pay-back times, additional BOS costs, and even raw material reserves. Last but not least, the co-evolution of a suitable energy storage network (e.g. hydrogen) is also foreseen to be a mandatory requirement. Cumulative installed capacity worldwide is projected to reach 9,000 GWp in 2050 in the most optimistic scenario, which is reduced to 2,400 GWp in the intermediate scenario. In the third "pessimistic" scenario the current economic incentives are not assumed to be sustained long enough to allow PV to become competitive with bulk electricity, resulting in a stunted market growth (500 GWp in 2050). The resulting predictions in terms of costs range from 0.50 to 1.50 €/Wp in 2050, respectively corresponding to 2 - 8 €-cents per kWh in Southern Europe and 4 - 14 €-cents per kWh in Northern Europe. Within the framework of these three general scenarios, special attention is then put to the role that is likely to be played by thin film technologies, namely amorphous Si, CdTe and CIS/CIGS. These technologies are expected to collectively reach a market share of approximately 45% by as early as 2025 in all but the most pessimistic scenario, wherein the same goal is put off until 2050. Marked increases in module efficiencies and material and energy consumption are also expected, to varying degrees depending on the assumptions made in the three scenarios.

INTRODUCTION

The rapid, exponential growth of photovoltaic cumulative installed capacity that has taken place in the last decade in Europe, Japan and the USA (the three largest world markets for PV) is illustrated in Figure 1. Four types of installations can be identified, i.e.: grid-connected centralized (large power plants); grid-connected distributed (smaller rooftop and façade systems); off-grid non-domestic (power plants and industrial installations in remote areas); off-grid domestic (mainly stand-alone rooftop systems for houses in remote areas). These four types of installations greatly differ in their requirement for Balance of System (BOS) components, the main discriminating aspect being the requirement for battery storage in the off-grid systems.

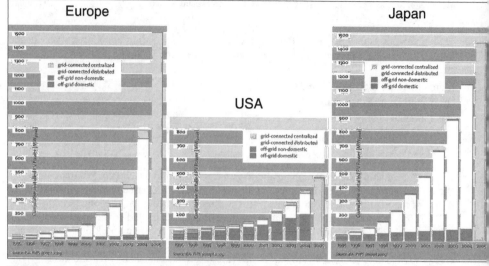

Figure 1. Cumulative installed capacity of PV systems [1].

Figure 1 also shows that the large majority of the recent installations in Europe and Japan is of the grid-connected distributed type. In the USA, on the other hand, more than half of the total installed power is still represented by off-grid systems (mainly stand-alone rooftop systems for use in remote areas). This can be at least in part explained by the lower population density of the United States, where PV has often been used as the most practical means of providing remote houses and towns with electricity, rather than a way to reduce the consumption of fossil fuels.

With growing concern regarding the global warming phenomenon and the associated environmental problems it brings about, as well as steadily increasing fuel prices and the controversial issue of energy security, renewable energy options are gathering more and more attention, and stimulating a lot of expectations for the future. In particular, photovoltaic electricity is increasingly being looked upon as a quintessentially "green" energy option for the future, since it entails virtually no emissions during the use phase, and large energy returns on investment (calculated EPBT for state-of-the-art PV systems range from approximately 3 to as little as 1 years [2-5].

However, in spite of their rapid growth in recent years, PV systems today still suffer from some non-negligible limits which have historically hampered their success, including comparatively high capital costs, fluctuating power output, and also often low social acceptability because of perceived reliability issues and poor aesthetics. It is therefore of wide-ranging interest to try and foresee what changes the future might bring in the PV sector.

The authors and colleagues at Ambiente Italia are leading the work package dedicated to photovoltaic technologies within the framework of the EU FP6 Integrated Project "New Energy Externalities Developments for Sustainability" (NEEDS), which is aimed at assessing the full costs (i.e. the direct manufacturing costs plus the indirect "external" costs based on Life Cycle Analysis) of all major future energy technologies, within the time horizon of the year 2050. The project is expected to have major policy relevance and to be used as a reference source in the upcoming years. It was recognized that four main interrelated aspects will mostly influence the diffusion scenarios of PV systems: 1) the technology development pathway; 2) the evolution of the direct costs; 3) the ongoing energy policies; and 4) the life cycle impacts and associated external costs. A wide-ranging analysis was therefore attempted which takes into consideration all four points, in order to assess the possible future success and overall environmental performance of solar electricity.

THE METHOD

An innovative method of analysis has been developed by the authors, which aims at integrating the traditional LCA approach with prospective methods typically used to reflect technological change (e.g. Technology Foresight and Experience Curves). In this way, "Environmental Learning Curves" can be identified, which express the variation of specific environmental LCIA indicators as a function of the expected installed capacity. The analysis is then envisaged to be extended to also estimate the indirect "external" monetary costs associated to environmental damage (currently under way).

Overall, the analysis is comprised of the following four steps:
1.   technology diffusion scenario analysis (three alternative scenarios as discussed below);
2.   technology development path (main factors affecting technological development are identified, and possible technology shifts are considered);
3.   parametric LCA for each scenario (with and without accounting for changes in background system);
4.   estimate of external costs (environmental damage costs associated to selected LCIA categories, still under way).

In particular, within step 2, the following key life cycle parameters were singled out for all the investigated technologies:
-    module efficiency;
-    module lifetime;
-    material resource consumption (with special reference to technology-specific materials);
-    energy resource consumption (both in terms of kWhel/Wp and kg(fuel)/Wp);
-    production of non-recyclable waste (with special reference to hazardous and radioactive waste);
-    Balance Of System performance ratio (in function of specific application and location).

The technology specification matrices resulting from step 3 were then sent out for review and were approved of by external experts (including EPIA, PV companies and selected researchers).

Finally, while the fourth step of the analysis is still under way, some preliminary results are already available for selected technologies and will be presented here.

SCENARIOS

In trying to draft possible future scenarios for PV, both in terms of costs and of environmental performance, the following key factors had to be considered:
-    Cost reduction.   PV costs have been declining steadily over the last two decades, with an average Progress Ratio of 80% (i.e. a cost reduction of 20% every doubling of production). It is widely agreed that in order for a similar trend to be maintained long enough for

PV electricity to eventually become economically competitive with utility peak power, it is mandatory that the current subsidizing schemes be maintained for at least one more decade. Thin film technologies such as CdTe benefit from the clear advantage of requiring far smaller amounts of photoactive materials than c-Si technologies, and cost reductions in the order of -75% are envisaged for these technologies [1], to the point of eventually reaching below 0.5 US$/Wp [6].

- Efficiency increase. Even if the energy return on investment of photovoltaic electricity is already quite good, a further efficiency increase is deemed to be feasible for all PV technologies, as illustrated in Figure 2. An authoritative study by the New Energy and Industrial Technology Development Organization [7] reports a target module efficiency of 22% for CIS/CIGS modules by 2030, and based on past trends and the current state of the art of the two technologies, similar performance is also likely to be attained by CdTe at the same time or shortly afterwards.

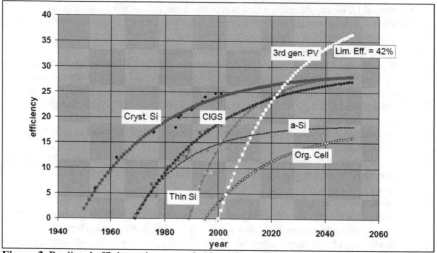

Figure 2. Predicted efficiency increases for the different PV technologies [8]

- Building integration. Enhanced structural integration of PV systems in new buildings as well as during restoration and/or renewal of older buildings will significantly contribute to the reduction of the BOS energy (and monetary) costs in all decentralized installations. Building integration of PV on a large scale also raises the further issue of social acceptability. This is especially important in the case of historically relevant sites and city centres, the likes of which are often encountered in Europe. However, innovative design and engineering solutions are being developed in order to facilitate the visual integration of PV into existing buildings, even including historical monuments. The authors had direct experience of the effectiveness of such design efforts during their previous involvement in the EU FP4 project PVACCEPT, which featured demonstration objects consisting of screen-printed thin film modules installed on historical buildings in northern Italy and Germany.

- Storage network. Integration of PV with large energy storage systems will be mandatory in order to warrant the necessary stability of the network if PV is ever to provide more than 10% of the total electricity supply of a country. One option that is currently being considered in this sense is represented by electrolytically produced hydrogen gas. The latter could be used as an energy buffer whereby to store the surplus energy generated by PV systems during peak irradiation hours, only to be converted back to electricity by means of fuel cell devices when the need arises. Other available energy storage options are pumped hydroelectric and compressed air energy storage (CAES); progress is also being made in the development of efficient high-speed flywheel systems whereby electric energy is converted into kinetic energy in a cylindrical or ringed mass, levitated by magnets and spinning at very high speeds (~10,000-20,000 rpm) in a vacuum chamber.

All these issues were taken into due consideration in drafting the following three alternative scenarios for the future development of the PV sector up to 2050.

1.   "Pessimistic" scenario.   This first scenario essentially mirrors the "best case" scenario drafted by IEA and OECD in their "Energy Technology Perspectives 2006" report [9], wherein it is assumed that PV will at best cumulatively account for approximately 2% of the overall world electricity supply by 2050 (the latter being estimated by IEA at 35,000 TWh/a). This comparatively pessimistic scenario essentially corresponds to assuming that the current incentives for PV will not be supported long enough for the technology to ever become competitive with bulk electricity. The gains in module efficiency are also expected to be slow, with both c-Si and CdTe/CIS/CIGS struggling to improve significantly upon their current levels of performance by 2025, and eventually only reaching 18% efficiency by 2050. Of course, this prediction reflects low R&D funds likely to be invested in these technologies in the event that they are not supported long enough for them to become economically competitive on a large scale. In particular, a very slow growth is foreseen for thin film PV, the market share of which only increases, in this scenario, to 15% in 2025, and eventually reaches 45% no sooner than 2050.

2.   "Optimistic / Realistic" scenario.   In this intermediate scenario, the predictions for the growth of the world PV market made by the European Photovoltaic Industry Association together with Greenpeace in their latest Solar Generation Report [1] are assumed to be valid all the way through to 2025, when the annual installed capacity is expected to reach 55 GWp. After that date, a transition is assumed to a less steep annual growth rate, eventually leading to a linear trend, whereby the cumulative installed capacity will keep growing steadily, approximately doubling each decade to eventually reach 2,400 GWp in 2050. This latter assumption is in good accordance with the predictions made in the latest report by the European Renewable Energy Council [10]. In this scenario, it is assumed that c-Si and thin film technologies will co-exist all along, with "third generation" novel devices (including low-cost devices such as organic-based PV and very-high-efficiency devices such as quantum cell and solar concentrator systems) only starting to gain ground after 2025. Thin film technologies will instead start expanding much sooner, already growing to approximately 45% of the market by 2025, with larger contributions by CIS/CIGS and CdTe and a gradual decline of a-Si.

3.   "Very Optimistic Scenario".   In this third scenario, bold annual growth rates are assumed from as early as 2010, and the trend is expected to keep growing in a quadratic fashion all the way through, topping out at almost 9,000 GWp in 2050. What this growth scenario implies is that by the mid-2030's at the very latest a large-scale energy storage infrastructure will have to have been developed. This last scenario is also dominated by the predicted very rapid expansion of PV systems based on novel devices after 2025 (following what can be referred to as

a major "technological breakthrough"). These novel technologies are expected to grow as much as to eventually account for approximately 50% of the total PV market in 2050.

The overall PV market growth according to the three scenarios is illustrated in Figure 3, with superimposed the average market growth rate for each decade expressed as percentage.

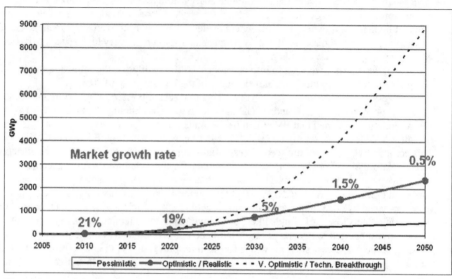

Figure 3. Expected PV market growth and market growth rates to 2050.

Table 1 then lists the different expected market shares of the three "families" of PV technologies (i.e. c-Si, thin films and novel devices) according to the three scenarios.

Table 1. Expected market shares of the different PV technologies

VERY OPTIMISTIC	% share		
	Present (2006)	2025	2050
c-Si	90%	50%	15%
Thin Films	10%	45%	35%
Novel Devices	0%	5%	50%
OPTIMISTIC / REALISTIC	**% share**		
	Present (2006)	2025	2050
c-Si	90%	50%	35%
Thin Films	10%	45%	35%
Novel Devices	0%	5%	30%
PESSIMISTIC	**% share**		
	Present (2006)	2025	2050
c-Si	90%	85%	50%
Thin Films	10%	15%	45%
Novel Devices	0%	0%	5%

RESULTS

Global Warming Potential

Figure 4 shows a comparison of the predicted Global Warming Potential of building integrated ribbon-Si and CdTe PV, as well as of future solar concentrator systems based on III/V semiconductors, according to the "Optimistic / Realistic" scenario discussed above (these preliminary results do not yet take into account the projected change in background data, i.e. the specific GWPs of the input flows are kept constant at the present level).

In order to put these numbers into perspective, it should be noted that the present performance in terms of Global Warming Potential of coal-fired power plants and of natural gas combined cycle plants are respectively about 900 and 400 g CO_2/kWh; the European electricity mix (UCTE mix) stands in between at 454 g CO_2/kWh. It can thus be seen that the expected GWP of thin film PV electricity in 2025 is expected to be no less than two orders of magnitude lower than that of the current European mix, a remarkable result that in the opinion of the authors ought not to be underestimated when making policy decisions about energy supplies for the future.

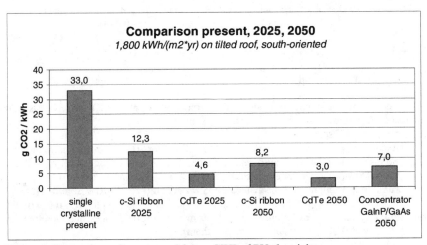

Figure 4. Comparison of present and future GWP of PV electricity.

Figure 5 illustrates a further interesting comparison in terms of GWP, i.e. that between PV electricity and the current status of selected non-fossil fuel electricity sources such as wind, hydropower and nuclear. Multi-crystalline Si is chosen as the most representative PV technology for the present, and an average of the calculated results for 2050 in the "Optimistic / Realistic" scenario is employed as expected future performance. The relevant indication here is the dramatic improvement that PV is expected to have, to an extent which appears to be less likely achievable by more mature technologies such as hydro or nuclear.

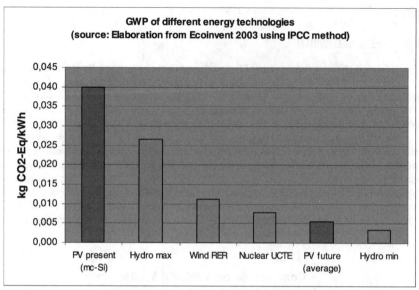

Figure 5. Comparison of GWPs of PV and other non-fossil-fuel electricity.

Capital Costs and External Costs

The different assumptions made for the purposes of estimating PV capital costs all the way to 2050 according to the three scenarios are summarized below.

1)   "Pessimistic" scenario:
-       fixed learning rate for PV modules = 20% / 10% after 2025 (in this scenario there will be little to no market penetration of third generation devices, hence the reduced LR after 2025);
-       variable learning rate for electrical BOS = 20% until 2010 / 5% from 2011;
-       variable learning rate for mechanical BOS = 20% until 2010 / 10% from 2011;
-       fixed allocation of mechanical BOS to PV for building integrated PV = 100% (in this road map PV systems will essentially remain after-market add-on devices, and therefore their entire mechanical BOS will remain allocated to them).

2)   "Optimistic / Realistic" scenario:
-       fixed learning rate for PV modules = 20% (the assumption of such a sustained LR is consistent with the foreseen market penetration of thin films after 2010, and then with the major technological shift to third generation devices after 2025);
-       variable learning rate for electrical BOS = 20% until 2010 / 10% 2011-2025 / 5% after 2025;
-       variable learning rate for mechanical BOS = 20% until 2010 / 10% from 2011;
-       variable allocation of mechanical BOS to PV for building integrated PV: 100% until 2010, then -1% each year to 85% in 2025; fixed at 85% afterwards (like in road map n.2, PV will become more and more a standard component of buildings, however the sheer bulk of the installations will be inferior, hence the more limited reduction of the allocation factors).

3)   "Very Optimistic" scenario:

- fixed learning rate for PV modules = 20% (the assumption of such a sustained LR is consistent with the foreseen market penetration of thin films after 2010, and then with the major technological shift to third generation devices after 2025);
- variable learning rate for electrical BOS = 20% until 2010 / 10% from 2011 ;
- variable learning rate for mechanical BOS = 20% until 2010 / 10% from 2011 ;
- variable allocation of mechanical BOS to PV for building integrated PV: 100% until 2010, then -2% each year to 20% in 2050 (BIPV will become more and more a standard component of buildings, and therefore a larger and larger share of the costs of the associated mechanical structure is to be attributed to the buildings themselves).

Figure 6 shows the results of these calculations for building integrated installations (which are envisaged to be the most important type) according to the three scenarios. A similar curve was also calculated for centralized plant-size installations. Our results are in good accordance with those reported in the PV-TRAC study [11], as shown in the superimposed boxes.

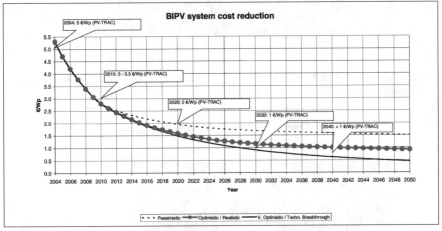

Figure 6. Expected evolution of the capital costs of building integrated PV systems

Some preliminary results can also be presented regarding the estimate of the expected average total cost of PV systems including the "external" cost of CO_2 mitigation, the latter having been somewhat conservatively estimated at 23 € per tonne of CO_2 in 2025 and 35€/t(CO_2) in 2050 [10]. In particular, it is interesting to compare the future total cost of PV electricity in the intermediate "Optimistic/Realistic" scenario to that of coal electricity, which is currently the "cheapest" fossil fuel option. In order to keep the playing field as level as possible, the following assumptions are made for future coal-fired power plants, again in accordance with EREC [10]: (a) the level of CO_2 emissions will be reduced to 728 g/kWh by 2025, and to 697 g/kWh by 2050; (b) the feedstock coal price will have risen to 72.9 US$/t in 2025 and to 86.4 US$/t in 2050. The results of this comparison are illustrated in Figures 7 and 8.

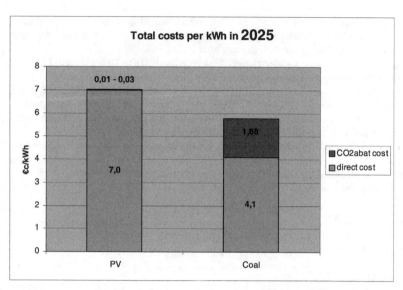

Figure 7. Projected total economic cost of PV and coal electricity including external cost for CO_2 abatement, in 2025.

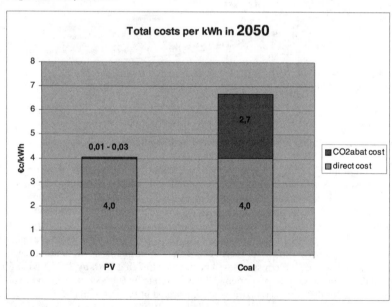

Figure 8. Projected total economic cost of PV and coal electricity including external cost for CO_2 abatement, in 2050.

A few preliminary considerations can be made. Firstly, the very low equivalent CO_2 emissions associated to the life cycle of photovoltaic technologies make the contribution of the external cost for CO_2 mitigation absolutely negligible. The same does not hold true for fossil fuel-based electricity, though, and the result is a marked competitive advantage for PV. Secondly, even in absolute terms, the overall total costs of PV electricity by 2050 are expected to be significantly lower than those for coal electricity, a remarkable achievement which should not be underestimated.

CONCLUSIONS

If economically supported through suitable incentives and feed-in tariffs for at least another decade, PV is likely to outgrow its current market niche and become a major player in the global electricity scene, significantly contributing to the lowering of the carbon intensity of future economies. Decentralization of bulk electricity production and availability of large energy storage networks are recognized as being two further necessary requisites for such large scale PV scenarios to become a reality.

If all these conditions are met, cumulative installed capacity may skyrocket, and Si-based PV will likely lose its preeminence to a widespread diffusion of thin film technologies (and possibly, at a later stage, third generation devices).

In any case, it is the authors' firm belief that in no circumstances is any single renewable energy technology likely to be able to effectively replace on its own the leading role that fossil fuels have had in the past century. Unfortunately, the ongoing debate that is often witnessed between the most fervent proponents of renewable energy technologies such as PV and those who systematically play them down as being almost irrelevant is counterproductive and completely misses the point. What will be necessary in order to meet the energy demand of future generations is instead a joint effort to make the best possible use of all available technologies, and in doing so their actual potential to be scaled up and their comparative associated environmental impact should always be taken into careful consideration.

REFERENCES

1. EPIA, 2006. "Solar Generation – Solar electricity for over one billion people and two million jobs by 2020". Greenpeace and European Photovoltaic Industry Association, The Netherlands / Belgium.
2. M. de Wild-Scholten and E. Alsema Environmental Life Cycle Inventory of Crystalline Silicon Photovoltaic Module Production Presented at Materials Research Society Fall 2005 Meeting, Boston, USA (2006)
3. E. Alsema and M. de Wild-Scholten, Environmental impacts of crystalline silicon photovoltaic module production. Presented at Materials Research Society Fall 2005 Meeting, Boston, USA (2006)
4. V. Fthenakis and C. Kim Energy Use and Greenhouse Gas Emissions in the Life Cycle of Thin Film CdTe Photovoltaics. Presented at Materials Research Society Fall 2005 Meeting, Boston, USA (2006)
5. Fthenakis V.M. and Alsema E., Photovoltaics Energy Payback Times, Greenhouse Gas Emissions and External Costs: 2004-early 2005 Status, Progress in Photovoltaics: Research and Applications, 14:275-280 (2006).

6. J.E. Trancick and K. Zweibel Technology choice and the cost reduction potential of photovoltaics. 4th World Conference on Photovoltaic Energy Conversion, Waikoloa, Hawaii, USA (2006)

7. NEDO, Overview of "PV Roadmap Toward 2030". New Energy and Industrial Technology Development Organization (NEDO), Kawasaki, Japan (2004)

8. Goetzberger, 2002. "Applied Solar Energy". Fraunhofer Institute for Solar Energy Systems (FhG/ISE), Germany (2006)

9. IEA/OECD, Energy Technology Perspectives 2006. Scenarios and Strategies to 2050. IEA Publications, Paris

10. EREC, "Energy [r]evolution - A sustainable world energy outlook. Global energy scenario report". Greenpeace and European Renewable Energy Council, The Netherlands (2007)

11. PV-TRAC, A Vision for Photovoltaic Technology. Photovoltaic Technology Research Advisory Council (PV-TRAC), EC (2005)

Mater. Res. Soc. Symp. Proc. Vol. 1041 © 2008 Materials Research Society   1041-R01-04

Comparative Life-cycle Analysis of Photovoltaics Based on Nano-materials: A Proposed Framework

V. Fthenakis[1,2], S. Gualtero[1], R. van der Meulen[1], and H. C. Kim[2]
[1]Center for Life Cycle Analysis, Columbia University, New York, NY, 10027
[2]PV Environmental Research Cener, Brookhaven National Laboratory, Upton, NY, 11973

ABSTRACT

Life cycle analysis is especially important for characterizing novel forms of material in new energy-generation technologies that are intended to replace or improve the current infrastructure of energy production. We propose a comparative life-cycle analysis framework for investigating the effect of incorporating nanotechnology in the life cycle of new photovoltaics, focusing on the differences between the new technologies and the ones that they may replace. Within this framework, we investigate the following parameters: Methods of synthesizing nanoparticles, physicochemical specifications of the precursors, material-utilization rates, deposition rates, energy-conversion efficiencies, and lifetime expectancy of the final product. We introduce the application of this framework by comparing nanostructured cadmium telluride and silicon films with their nano- and amorphous-structured equivalents.

1. INTRODUCTION

Research is rapidly expanding in academia and industry on synthesizing nanoparticle precursors and fabricating nanostructured solar cells from them. The drivers for this research include the quest for reducing losses due to photon reflection, increasing photon-to-electron conversion-efficiencies, lowering the costs of manufacturing, and enabling easier and cheaper applications/installations. While nanotechnology may meet one or more of these objectives, there is a dearth of knowledge about the environmental implications of using nanomaterials. Life Cycle Analysis (LCA) provides a framework for studying them. LCA is widely used to evaluate the environmental impacts from extracting and processing raw materials, and from manufacturing, using, and disposing of the end-of-life of products. This type of analysis is needed especially for characterizing material forms in promising energy-generation technologies intended to replace or improve the current energy-production infrastructure. The goal of this study is to examine the energy- and material-requirements and emissions in the life cycles of representative nanotechnology applications in solar-energy conversion, and compare them with the material- and energy- flows in the technologies that they may replace.

2. METHODOLOGY

We describe a life-cycle analysis framework that will allow us to compare the environmental impacts of nanomaterials and bulk materials during the production of photovoltaics (PVs). We will investigate the following parameters that distinguish nanotechnology from conventional technology: 1) Methods of synthesizing the nanoparticles; 2) physical specifications of the precursors; 3) material utilization rate/process efficiency; 4) deposition processes and parameters (deposition rate, temperature, pressure); 5) energy-conversion efficiency of the solar cells; and 6) life-time expectancy of the final product. Figure 1 illustrates the proposed conceptual framework for this comparable LCA.

We selected two paradigms for comparison: One is a future third-generation solar cell grown from a solution of nano-structured CdTe/CdSe precursors; the second is the near-commercialization option of adding a nano-crystalline silicon layer on amorphous Si configurations. The first method is based on entirely different processes than those employed in its micro –material equivalent, whereas the second involves only tuning the parameters in an existing deposition process. For inventory analyses of each of the two paradigms, we developed flow diagrams, and quantified the requirements for energy and materials, and the emissions.

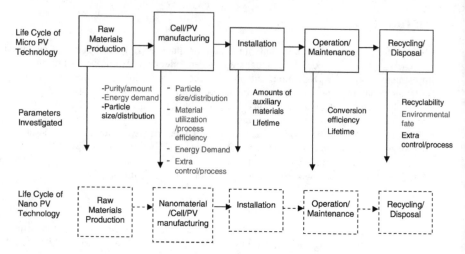

Figure 1. Framework for the Comparable LCA of Micro- and Nano-structured Photovoltaics

3. COMPARISON OF NANO- AND MICRO- CRYSTALLINE CdTe PV

Fthenakis and Kim (2006) described an LCA of commercial CdTe PV modules using data obtained from First Solar's production plant in Perrysburg, Ohio [1]. The facility produces frameless, double-glass, CdTe modules rated at 10% efficiency. The semiconductors are deposited by vapor transport deposition (VTD) that is based on the sublimation of CdTe and CdS powders and the subsequent condensation of the vapors on glass substrates.

Researchers at Lawrence Berkeley National Laboratory and the University of California in Berkeley have been formulating a solution-based process to produce the next generation of solar cells from CdTe- and CdSe-nanorods. So far, small-area solar cells with 2.9% conversion efficiency have been obtained [2]. Laboratory production involves synthesizing CdSe- and CdTe-nanorods and fabricating the device. For the synthesis of CdSe nanorods, CdO and phosphorus materials are mixed and degassed at 120 °C, and, after adding trioctylphosphine (TOP) and Se, the particles are left to grow; then anhydrous toluene and anhydrous isopropanol are added to precipitate CdSe nanocrystals. The CdTe nanocrystals are made in the same way by adding Te. For fabricating the CdSe/CdTe devices, the nanorods are dispersed in pyridine and stirred overnight under reflux. Next, they are precipitated with hexane, washed with toluene, and re-dissolved in pyridine. The solution is ultrasonicated for 30 minutes and filtered. Then, the nanocrystals are spin-cast on to glass substrate coated with indium tin oxide (ITO) and a layer of Al_2O_3, followed by spin casting of a solution of $CdCl_2$ in methanol; the samples are heated and held under pressure overnight. Finally, top electrodes are deposited on them by thermal evaporation through a shadow mask. This process produces individual devices with 0.03 cm^2 nominal area. Figure 2 schematically depicts the pathways.

1. Synthesis of Nanorods

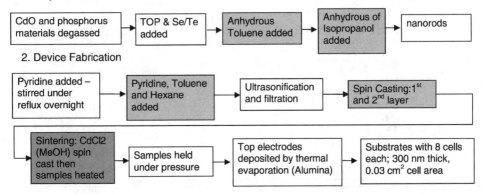

2. Device Fabrication

Figure 2. Nanostructured -CdTe fabrication processes (In gray: Major changes expected as the process is scaled up)

We undertook a preliminary mass balance of this laboratory process, based on the reported materials and amounts reported in synthesis. Table 1 lists the total quantities of the Cd and Te precursors and seven solvents used. We estimated that material utilization for synthesizing the nanorods, defined as the amount of CdO and Te or Se per mass of rods produced, was 73% for CdSe, and 77% for CdTe rods. The material used in fabricating the devices is extremely low; this is satisfactory in the laboratory where only very small area cells are needed. However, linear extrapolation of the material quantities from this scale to a device with an area of one m^2, results in an incredibly high 610 kg/m^2, which is a commercial impediment. Undoubtedly, the process must be optimized for scaling up to market levels, so material utilization most likely will improve in the future.

Formulating an LCA for emerging technologies which are still at the early stages of development imposes many challenges reflecting the uncertainty, and sometimes the unpredictability of the design of the final process. Hence, at this stage, the researcher's goal, understandably, is to make a working device. Issues such as material utilization usually are addressed later as the process is scaled-up. In this particular paradigm, we forecast the following outcomes: a) Material utilization can be greatly improved, especially for solvents, purifying agents, and surfactants; b) spin casting, the current solution-spreading technique, is inefficient for large-area depositions and likely will be replaced by a printing process. In spin casting, up to 99% of the solution is wasted, whereas with inkjet printing it is less than 2% [3]. We expect the latter process to be widely used for these types of applications. Nanosolar, of Palo Alto, California, announced that they developed a high-yield ink printing process for chalcopyrite

CIGS nanoparticles [4]; accordingly, we can assume that this technique also could be employed for forming very thin films of CdTe and CdSe. For a prospective analysis, the LCA uses two scenarios. The first, called "nano-Labscale", linearly scales up the current lab-scale material and energy requirements. The second scenario, termed "nano-Projected Commercial" is based on inkjet printing. Table 1 shows the differences in the material requirements between the two scales. As expected, linear scaling up of the laboratory process would be unsustainable as it would require relatively high volumes of the precursors Cd, Te, Se, and tremendous volumes of the solvents.

Table 1. Comparision of material requirements (g/m^2) in various technologies

Materials	Micro Commercial	Nano - Lab	Nano - Commercial
CdTe	47.7		
CdS	2.9		
Cd Comp.	0.8		
Metal Comp.	0.0		
CdO		1,111	11.2
Te		170	1.7
Se		203	2.0
Solvents		609,842	6,160
Phosphorus compounds		32,264	326

*Commercial CdTe PV does not use solvents; the total greenhouse gas emissions from the life cycle of a whole PV module correspond to ~15 g CO_2/m^2.

4. COMPARISON OF AMORPHOUS- AND NANOSTRUCTURED- SILICON PV

Several companies produce amorphous-silicon photovoltaics via plasma deposition from silane/hydrogen mixtures. Pacca et al [5] conducted LCA studies of such modules from United Solar, Troy, Michigan. The latest developments in amorphous-silicon PV are implementing a nano-structured silicon layer (nc-Si) to create a cell with a wider gap, so improving the device's efficiency. Major research efforts concentrate on two different designs: United Solar has replaced the intrinsic SiGe layer in its triple-junction cell with a nc-Si layer. Oerlikon has added a layer of μc -Si in its amorphous cells to produce a tandem a-Si/micro crystalline called "micromorph". We note that the terms nano-crystalline (nc-Si) and microcrystalline (-Si) silicon sometimes are used interchangeably as grain sizes range from a few nanometers to one micrometer. We based our reference case on the evolution of the amorphous silicon solar cell to the "micromorph" one. The latter is structured with an a-Si top layer of around 200- to 300-nm, and bottom layers of 1000- to 2000-nm thick, surrounded by glass substrate and encapsulated in glass (Figure 3). Similar to amorphous-silicon deposition, nc-Si layers are laid down by plasma-enhanced chemical vapor deposition from mixtures of silane and hydrogen, but, to induce the formation of the crystalline phase, the silane gas is highly diluted by hydrogen (e.g., H_2/SiH_4 ratio of 16).

The nc-Si layer improves the spectral response to light in the infrared range, and it not susceptible to light-induced changes, as the a-Si layers are. The electrical conversion efficiency is expected to increase from 7 % for a-Si, to 8.6 % for a-Si/nc-Si [6]. The forecasted efficiencies for 2015 are, respectively, 9% and 12% for an a-Si and a micromorph cell. A typical deposition cycle for a-Si and nc-Si consists of loading the substrate, creating a vacuum in the chamber, heating the chamber/substrate, depositing the film, and unloading it. The reactor must be cleaned between the deposition of every layer to avoid contamination. Oerlikon currently uses SF6 for such cleaning, which subsequently is recycled and abated on site [6].

The design of the cell and the required modifications in the deposition process lead to potentially higher energy requirements for an a-Si/nc-Si structure than an a-Si one. This greater demand is attributable to the following changes: a) Thicker layers with correspondingly longer deposition time; and, b) a higher excess of hydrogen. Oerlikon Solar deposits the a-Si layer at a rate of 0.3 nm/s, and the nc-Si layer at 0.5 – 0.6 nm/s [6]. Nanocrystalline silicon is formed using silane from a dilute silane-hydrogen mixture (5-7%) [7], whereas depositing amorphous silicon utilizes about three times less hydrogen (i.e., 16-20% silane-hydrogen mixture). The utilization rates of silane in the a-Si and nc-Si depositions are approximately the same, in the range of 20-30%.

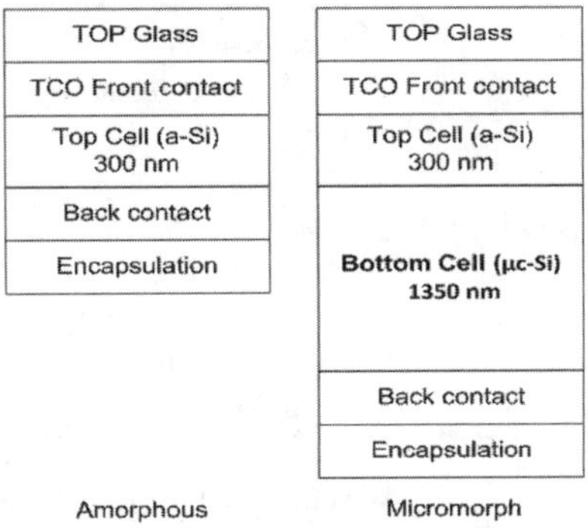

Figure 3. Cell structures; amorphous vs. IMT/Oerlikon micromorph concept

A major requirement for conducting a LCA is obtaining process material- and energy-data from a commercially mature operation. In the current a-Si/nc-Si paradigm, we can use the material- and energy-inventories of a-Si PV production with some modifications and additions. We started from LCAs of a-Si photovoltaics and investigated the environmental implications of adding nc-Si layers. The most recent publication is that by Pacca et al, 2006 [5] who describe production at United Solar, Auburn Hills, MI. However, their study gives only aggregated energy data, and it involves different substrates and encapsulants than those in the design considered by Oerlikon. Thus, we used detailed, disaggregated process-data from earlier studies [8, 9], and linearly adjusted them to reflect the improvement in total energy use, as shown in the 2006 study (i.e., 445 MJ/m2). The direct energy use in a-Si deposition mostly is from operating vacuum pumps and heating the substrates, both of which depend on deposition time. We estimated deposition time from the layer's thicknesses and the deposition rate. Table 2 lists the parameters defining the two fabrication lines, and Table 3 gives the sources of life- cycle inventory data for materials and energy inputs.

Table 2. Comparison of deposition process parameters

	Amorphous	Micromorphous
Thickness top layer (nm)	300	300
Thickness second layer (nm)	N.A.	1350 (1000-1500)
Deposition rate (nm/s) top layer	9	3.2
Deposition rate (nm/s) bottom layer	N.A.	5 (4-6)
Reactor cleaning cycles	1 (= 116 g/m^2 SF$_6$*)	2 (= 231 g/m^2 SF$_6$*)
Silane input (g/m^2)	2.8	15.4
Hydrogen input (g/m^2)	16.9	213

* Thermal abatement in SF6 emissions is assumed to be 99 % [10]

Table 3. Material and energy inventory data of frameless a-Si module per m^2 module area

Materials and Energy	Usage	LCI source
Flat glass, coated (kg)	14.64	Ecoinvent
Phosphine (g)	0.17	Lewis and Keoleian, 1997
Sulphur hexafluoride (g)	116	Ecoinvent
Oxygen (g)	1.1	BUWAL 250
Hydrogen (g)	16.9	BUWAL 250
Silane (g)	2.8	Lewis and Keoleian, 1997
Cell material (MJ)	10.2	
Module manufacturing (MJp)	445	Pacca, personal communication
Overhead (MJp)	250	Alsema, 1998; 2000
Total (MJp)	705	

We compared the differences in fabricating an a-Si single junction structure and an a-Si/nc-Si 'micromorph' in terms of energy payback times (EPBTs), and CO_2 emissions; both energy use and CO_2 emissions were significantly higher from the latter process. This is mainly due to the higher energy use, and higher consumption of hydrogen and SF_6. However, an a-Si/nc-Si cell will generate more electricity than a-Si cell because of the increased conversion efficiency of the former, thereby requiring a similar time to pay back the energy used during the production stage (Table 4).

Table 4. Preliminary results in terms of Energy Payback Time (EPBT) and Greenhouse Gas (GHG) emissions

	a-Si module	a-Si/μc-Si module
Module efficiencies		
Current [11]	7.0%	8.6%
Forecast for 2015 [11]	9.0%	12.0%
EPBT*		
Current (yr)	2.6	2.7
Forecast for 2015 (yr)	2.0	2.0
GHG Emissions		
CO_2 only, current (kg/m²)	59	74
Total GHG, current (kg CO_2-eq./m²)	94	141
CO_2 only, forecast for 2015 (kg/m²)	20	20
Total GHG, forecast for 2015 (kg CO_2-eq./m²)	31	38

*Without frame, assuming US average insolation (1800 kWh/m²/yr), a system performance ratio of 0.8

On the other hand, the improvement in efficiency expected by adding a nano-crystalline Si layer on an a-Si device does not offset the additional greenhouse gas (GHG) emissions related to this change. Disproportionally high amounts of GHGs are anticipated due to a) the increase of deposition times and corresponding usage of energy; b) increased requirements for cleaning the reactor, and SF_6 use; and, c) increased use of hydrogen as a more dilute SiH_4/H_2 mixture is needed. We note that the preliminary results presented here are indicative rather than informative because data from fully commercialized scale facilities are unavailable for an a-Si/nc-Si module, and the disaggregated data for an a-Si module may be outdated.

5. CONCLUSIONS

We propose a framework for assessing the environmental implications during the life cycles of nano-structured solar cells that utilizes the LCA of existing commercial PV, and focuses on the differences at each of the life-cycle stages of the nano- and the bulk-materials. Under two paradigms, we discussed a preliminary application of this framework to different technological options, identifying methodological issues in scaling-up laboratory processes and forecasting material and energy inventories. We consider that the proposed framework will contribute to reduce uncertainties in the environmental appraisal of nanotechnology by using validated data from current commercial processes and extending them according to the identified crucial processes and environmental parameters.

ACKNOWLEDGMENTS

This research is funded by the EPA STAR program under grant # RD-8333340. We are grateful to colleagues and researchers who kindly provided essential data for this study including Ilan Gur, UC Berkeley; Sergio Pacca, University of Michigan; Ulrich Kroll, Oerlikon Solar Lab; Christoph Ballif and Andrea Feltrin, Institute of MicroTechnology.

REFERENCES

1. V.M. Fthenakis and H.C. Kim, Energy Use and Greenhouse Gas Emissions in the Life Cycle of CdTe Photovoltaics. Material Research Society Fall Meeting, Symposium G: Life Cycle Analysis Tools for "Green" Materials and Process Selection, Boston, MA, 2005.
2. I. Gur, N.A. Fromer, M.L. Geiger, and A.P. Alivisatos, Science 310, 462 (2005).
3. J. Bharathan and Y. Yang, Appl. Phys. Lett. 72, 21 (1998).
4. Nanosolar website: http://www.nanosolar.com/printsemi.htm.
5. S. Pacca, D. Sivaraman, and G.A. Keoleian, Life Cycle Assessment of the 33 kW Photovoltaic System on the Dana Building at the University of Michigan: Thin film Laminates, Multi-crystalline Modules, and Balance of System Components. Center for Sustainable Systems, 2006. CSS05-09.
6. S. Benagli, J. H., D. Borello, J. Spitznagel, U. Kroll, J. Meijer, E. Vallat-Sauvain, H. Schmidt, G. Monteduro, B. Dehbozorgi, P.-A. Madliger, D. Zimin, O. Kluth, G. Buechel, A. Zindel, and D. Koch-Ospelt, High performance LPCVD-ZnO appleid in amorphous silicon single junction P-I-N and micromorph tandem solar device prepared in industrial KAITM -M R&D reactor. 22nd European Photovoltaic Solar Energy Conference & Exhibition. Milan, Oerlikon Solar-Lab S.A, 2007.
7. B. Strahm, A.A. Howling, L. Sansonnens, and C. Hollenstein, Plasma Sources Sci. Technol. 16, 80-89 (2007).
8. E.A. Alsema, Prog. Photovoltaics. 8, 17-25 (2000).

9. G.M. Lewis and G.A. Keoleian, Life cycle design of amorphous silicon photovoltaic modules. National Risk Management Research Laboratory. Office of research and Development. US EPA, 1997. EPA/600/R-97/081.

10. M. deWild-Scholten, E. Alsema, V. Fthenakis, G. Agostinelli, H. Dekkers, K. Roth, and V. Kinzig, Fluorinated Greenhouse Gases in Photovoltaic Module Manufacturing: Potential Emissions and Abatement Strategies. 22nd European Photovoltaic Solar Energy Conference, Milan, Italy, 3-7 September 2007.

11. F. Baumgartner, Future and further development of silicon thin film technology: prospects for R&D. Hochschule für Technik Buchs, NTB, Switzerland, 2007.

Mater. Res. Soc. Symp. Proc. Vol. 1041 © 2008 Materials Research Society

Life Cycle Assessment of Photovoltaics: Update of the ecoinvent Database

Niels Jungbluth[1], Roberto Dones[2], and Rolf Frischknecht[3]
[1]Head Quarter, ESU-services Ltd., Kanzleistr. 4, Uster, 8610, Switzerland
[2]Paul Scherrer Institut, Villigen PSI, 5232, Switzerland
[3]ecoinvent Centre, Dübendorf, Switzerland

ABSTRACT

Recently, the data for photovoltaics in the ecoinvent database have been updated on behalf of the European Photovoltaics Industry Association and the Swiss Federal Authority for Energy. Data have been collected in this project directly from manufacturers and were provided by other research projects. LCA studies from different authors are considered for the assessment. The information is used to elaborate a life cycle inventory from cradle to grave for the PV electricity production in grid-connected 3 kWp plants in the year 2005.

The inventories cover mono- and polycrystalline cells, amorphous and ribbon-silicon, CdTe and CIS thin film cells. Environmental impacts due to the infrastructure for all production stages and the effluents from wafer production are also considered. The ecoinvent database is used as background database.

Results from the LCA study are presented, comparing different types of cells used in Switzerland and analysing also the electricity production in a range of different countries. It is also discussed how the environmental impacts of photovoltaics have been reduced over the last 15 years, using the CED indicator. The consistent and coherent LCI datasets for basic processes make it easier to perform LCA studies, and increase the credibility and acceptance of the life cycle results. The content of the PV LCI datasets is available via the website www.ecoinvent.org for ecoinvent members.

INTRODUCTION

Life cycle assessment (LCA) has proved to be a powerful tool for the environmental improvement of production processes in the energy sector. However, the increased use of the LCA method to analyse systems is hindered by the lack of agreement on the use of methods and by the limited availability of life cycle inventory (LCI) data. Recently, the data for photovoltaics in the ecoinvent database have been updated on behalf of the European Photovoltaics Industry Association and the Swiss Federal Authority for Energy ([1]).

In the past years the PV sector developed rapidly. Ongoing projects such as *CrystalClear*[1] have investigated the up-to-date life cycle inventory data of the multi- and singlecrystalline technologies ([2]). Updated LCI data of single- and multicrystalline PV technologies were investigated within the framework of the CrystalClear project based on

---

[1] See www.ipcrystalclear.info for detailed information.

questionnaires sent to different involved industries. The data investigated with 11 European and US photovoltaic companies for the reference year 2005 are now implemented in the ecoinvent database v2.0 and documented according to the ecoinvent requirements ([1]). The following unit process raw data have been investigated and updated:

- multicrystalline SoG-silicon, Siemens process (new solar-grade process)
- multicrystalline-Si wafer (mc-Si or multi-Si)
- singlecrystalline-Si wafer (sc-Si or single-Si)
- ribbon Si wafer (so far not covered by ecoinvent data v1.3)
- ribbon-, multi- or single-Si cell (156 mm x156 mm)
- modules, ribbon-Si (new) and other module types
- silica carbide (SiC)
- PV-electricity mix Switzerland and in other countries
- recycling of sawing slurry and provision of SiC and glycol
- front metallization paste and back side metallization paste of solar cells
- inverter including electronic components

New thin film cells technologies like CIS or CdTe are entering the market. For the first time also thin film photovoltaics (CIS, CdTe and amorphous silicon) are investigated for the ecoinvent data based on literature information.

The yield per kW_p is one important factor for the comparison of PV with other types of electricity production. For ecoinvent data v1.3 only the situation in Switzerland had been investigated [3]. For the ecoinvent data v2.0 we investigated the PV technology mixes for several European countries using the specific electricity yields in each country based on published irradiation levels ([4]). Also yields in selected non-European countries (e.g. in Asia, Australia and North-America) were considered for a rough extrapolation of the European PV model to PV installations in those countries. However, different electricity/energy mixes for the manufacturing upstream chains have not been modelled for different country-specific cases but only the average European chain was investigated in detail.

SYSTEM BOUNDARIES

Sixteen different, grid-connected photovoltaic systems were studied. These are different small-scale plants of $3kW_p$ capacity and operational in the year 2005 in Switzerland (see Tab. 1).

The plants differ according to the cell type (single- and multicrystalline silicon, ribbon-silicon, thin film cells with CdTe and CIS), and the place of installation (slanted roof, flat roof and façade). Slanted roof and façade systems are further distinguished according to the kind of installation (building integrated i.e. frameless laminate, or mounted i.e. framed panel).

The average PV electricity mix in Switzerland considers the actual performance of the installed plants ([5]), while data for specific types of PV plants (e.g. laminate and panel, single- or multicrystalline) can be used for comparisons of different technologies. The calculations in ([4]) are performed for PV plants located in Berne with an annual yield of 922 and 620 kWh/kW_p

for roof-top and façade installations, respectively. This yield is calculated with an irradiation of 1117 kWh per m^2 and a performance ratio of 0.75. These results have been used for our technology specific assessments (e.g. in Fig. 3).

The actual PV electricity mixes in 2005 with different types of PV power plants in several countries are also modelled. The yield data for PV electricity mixes in other countries are based on a publication for optimum installations ([4]) and a correction factor which takes into account an actually lower yield of average installations in Switzerland compared to this optimum installation ([1]).

Tab. 1: Overview of the types of photovoltaic 3 kWp systems investigated for an installation in Switzerland

Installation	Cell type	Panel type [1]
Slanted roof	sc-Si	Panel
	mc-Si	Panel
	a-Si	Panel
	ribbon-Si	Panel
	CdTe	Panel
	CIS	Panel
	sc-Si	Laminate
	mc-Si	Laminate
	a-Si	Laminate
	ribbon-Si	Laminate
Flat roof	sc-Si	Panel
	mc-Si	Panel
Façade	sc-Si	Panel
	mc-Si	Panel
	sc-Si	Laminate
	mc-Si	Laminate

1) Panel = mounted; Laminate = integrated in the roof construction, sc-Si = singlecrystalline silicon, mc-Si = multicrystalline silicon.

The amount of panels necessary for a 3 kW$_p$ plant is calculated with the cell efficiency and the cell surface of the panel. The surface areas for a 3 kWp-plant are shown in Tab. 2. For a-Si and CIS there is no "cell" as such. Thus, the area of cell and panel is the same. Also the efficiency is not differentiated. Thus, it is the same for cell and panel.

Tab. 2: Active panel area of 3 kWp-PV plants with different types of solar cells, cell efficiencies and calculated panel capacity, amount of panels per 3kW$_p$ plant

cell type	cell efficiency	panel efficiency	cell area	cells	amount of panels per 3 kWp	active surface	panel capacity rate
	%	%	cm2	unit/m2	m2	m2	Wp/m2
sc-Si	15.3%	14.0%	243	37.6	21.4	19.6	140
mc-Si	14.4%	13.2%	243	37.6	22.8	20.8	132
ribbon-Si	13.1%	12.0%	243	37.6	25.0	22.9	120
a-Si	6.5%	6.5%	10000	1	46.5	46.5	65
CIS	10.7%	10.7%	10000	1	28.1	28.1	107
CdTe	7.6%	7.1%	9306	1	42.2	39.2	71

*a 7.1% CdTe panel efficiency assumes a 90% contribution of Antec (6.9% efficient) modules and 10% of First Solar (9%% efficient) modules in Switzerland in 2005.

All subsystems shown in Fig. 1 are included as individual datasets within the system boundaries for silicon based PV power plants. The process data include quartz reduction, silicon purification, wafer, panel and laminate production, manufacturing of inverter, mounting, cabling, infrastructure, assuming 30 years operational lifetime for the plant. The basic assumptions for each of these unit processes are described in the report. We considered the following items for each production stages as far as data were available:

- energy consumption,
- air- and waterborne process-specific pollutants at all production stages,
- materials, auxiliary chemicals, etc.
- transport of materials, of energy carriers, of semi-finished products and of the complete power plant,
- waste treatment processes for production wastes,
- dismantling of all components,
- infrastructure for all production facilities with its land use.

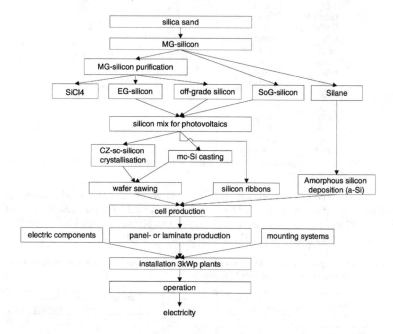

Fig. 1:   Different sub systems investigated for the production chain of silicon cells based photovoltaic power plants installed in Switzerland. MG-silicon: metallurgical grade silicon, EG-silicon: electronic grade silicon, SoG-silicon: solar-grade silicon, a-Si: amorphous silicon

All subsystems shown in Fig. 2 are included within the system boundaries for thin film PV power plants. All inputs (semiconductor metals, panel materials and auxiliary materials) for the production of thin film cells, laminates and panels are investigated in other reports of the ecoinvent project ([6]). Thus, in the specific report for PV we only described the process stages starting from the laminate and panel production.

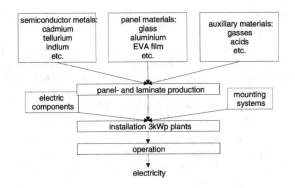

Fig. 2: Different sub systems investigated for thin film (CIS and CdTe) photovoltaic power plants installed in Switzerland

RESULTS

Pay-back time

An important yardstick for the assessment of renewable energy systems is the estimation of the energy and/or environmental pay back time. The outcome of such a comparison is influenced by the choice of the reference system on the one hand and the indicator on the other. Here we consider the UCTE electricity mix in year 2004 [7] as the reference system. Fig. 3 shows the pay-back-time for the non-renewable cumulative energy demand for PV power plants operated in Switzerland. This time is between 2.5 and 4.9 years for the different types of PV plants. Thus, it is 5 to 10 times shorter than the expected lifetime of the photovoltaic power plants. Different characteristics like type of installation, type of cells, type of panel (mounted, on Fig. 3) or laminates (integrated) are the key factors for determining the relative differences in results illustrated in this figure.

Fthenakis and Alsema have earlier reported pay-back times of 2.2 and 1 year for multi-Si and CdTe panels, respectively, under an irradiation of 1700 kWh/m2yr, which is the South European average [8]. Adjusted to the irradiation level of 1117 kWh/m2/yr in Switzerland this would correspond to 3.3 year for multi-Si and 1.5 year for CdTe.

When we compare the corrected EPBT values from Fthenakis and Alsema for multi-Si (3.3 yr) with our results for "slanted roof, multi-Si, laminated, integrated" systems (2.7 yr) we see that our result is lower. Reasons for this difference are the technology improvements in multi-Si module production since the analysis by Fthenakis and Alsema (e.g. reduced silicon consumption), and that we assumed a 7.1% efficiency for CdTe PV, whereas Fthenakis and Alsema used a 9% efficiency. Further analysis shows that energy requirements for the BOS are considerably higher in our study, for a number of reasons (heavier mounting structures, detailed estimate for inverter electronics).

Furthermore, the share of BOS for the total results in [8] is much lower than in our study and the difference between CdTe and multi-Si panels is only small. CdTe panels have in our study only 60% of the efficiency compared to multi-Si, which result in much higher specific share of mounting structures for the thin-film. In our study the BOS of CdTe amounts to 44% while [8] gives only a share of about 19%.

Comparison of the EPBT values for CdTe technology between Fthenakis and Alsema (1.5 yr), on the one hand, and our results on the other hand (2.7 yr, see fig. 3) is rather difficult because of differences in the considered module production technology, data sources, and module efficiency (9% vs. 7.1%). Tentatively, we could conclude that our higher energy requirements for the mounting structure specific to Switzerland, in combination with lower module efficiencies, explains a large part of our higher EPBT result for thin film PVs.

We caution that ecoinvent data are a careful analysis of the situation in Switzerland for the year 2005. They do not represent a reference for the current state-of-the-art of thin film PV mix in other countries, and should not be casually extrapolated to other situations or boundary conditions.

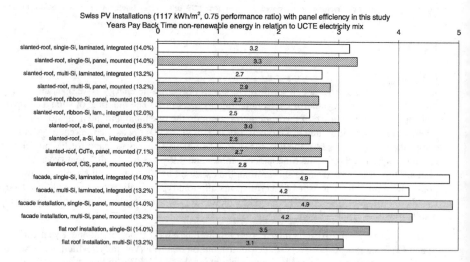

Fig. 3: Energy pay back time of 3 kWp photovoltaic power plants operated in Switzerland in relation to the UCTE electricity mix (results with ecoinvent data v2.0). Red for slanted roof, yellow for façade, blue for flat roof. Efficiencies in this, are shown in the captions above.

Development of LCA results

Fig. 4 shows the development of results for the cumulative energy demand of photovoltaic electricity in this study compared to previous Swiss studies. The figure shows also the increase in installed capacity in Switzerland. This evaluation shows that the cumulative energy demand has been decreased by a factor of 3 or more since the first studies on PV systems made in the early nineties.

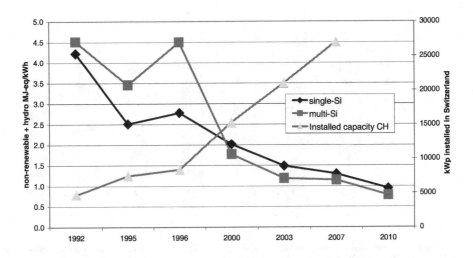

Fig. 4   Cumulative energy demand of the life cycle inventory for photovoltaic power production in this study (2007) and comparison with previous Swiss studies. Data for 2010 were forecasted with ecoinvent data v1.0 in 2003 [3, 9-11]

Comparison of different countries

Fig. 5 shows the global warming potential (100a) for photovoltaic power plants operated in different countries. The comparison shows that there might be considerable differences between different countries depending on the irradiation and thus on the actual yield per kW_p installed. CO_2-equivalent emission per kWh might be as low as 50 grams per kWh to the grid in the average case investigated for Spain. They will be even lower if optimum installations with best performance ratios are taken into account.

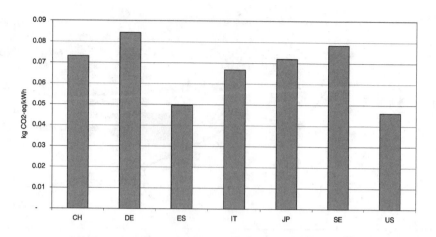

Fig. 5 Global warming potential in kg CO_2-eq per kWh for the average (2005) photovoltaic electricity mixes in different countries (results with ecoinvent data v2.0)

CONCLUSIONS

The life cycle inventories of photovoltaic power plants performed for the ecoinvent data v2.0 can be assumed to be representative for photovoltaic plants and for the average photovoltaic mix in Switzerland and in other European countries in the year 2005. The analysis of the results shows that it is quite important to take the real market situation (raw material supply, electricity, etc.) into account. All major PV technologies have been investigated in this study in a consistent and comparable way.

Differences in comparison to other studies are mainly due to different data sources, and assumptions for irradiation, efficiencies, and mounting structures. These factors must also be taken into account along with the technology development level for comparisons with other types of electricity generation. Other factors like differences in the share of imports from different PV producing regions or types of PV cells have not been modelled separately for each country. It should be considered that the inventory may not be valid for wafers and panels produced outside of Europe or the US, because production technologies and power mix for production processes are generally not the same. The datasets on PV electricity in non-European countries should thus be revised as soon as data are available for production patterns in major producing countries (e.g. Japan).

Our study shows that also the balance of system components play a more and more important role for the comparison of different types of PV technologies with different efficiencies and thus different sizes of mounting systems for the same electric output. The low efficiency systems need larger amounts of mounting structure and cabling which partly outweighs the better performance per kWp of module alone.

ACKNOWLEDGMENT

The research work on photovoltaics within the ecoinvent v2.0 project was financed by the Swiss Federal Office of Energy and the European Photovoltaic Industry Association (EPIA). These contributions are highly acknowledged.

Mariska de Wild-Scholten and Erik Alsema provided us the data from the CrystalClear project. But, besides they send many interesting further information and helped for discussing the appropriate data for different PV technologies. Furthermore they contributed detailed comments to first drafts of the final report. Thank you to both of you.

REFERENCES

1. Jungbluth N & Tuchschmid M, *Photovoltaics*, in *Sachbilanzen von Energiesystemen: Grundlagen für den ökologischen Vergleich von Energiesystemen und den Einbezug von Energiesystemen in Ökobilanzen für die Schweiz*, Dones R, (Editor). 2007, Paul Scherrer Institut Villigen, Swiss Centre for Life Cycle Inventories: Dübendorf, CH. p. 180. www.ecoinvent.org.

2. de Wild-Scholten MJ & Alsema EA. *Environmental Life Cycle Inventory of Crystalline Silicon Photovoltaic Module Production*. in *Proceedings of the Materials Research Society Fall 2005 Meeting*. 2005. Society MR, Boston, USA. Retrieved from www.mrs.org.

3. Jungbluth N, *Photovoltaik*, in *Sachbilanzen von Energiesystemen: Grundlagen für den ökologischen Vergleich von Energiesystemen und den Einbezug von Energiesystemen in Ökobilanzen für die Schweiz*, Dones R, (Editor). 2003, Final report ecoinvent 2000 No. 6-XII, Paul Scherrer Institut Villigen, Swiss Centre for Life Cycle Inventories: Dübendorf, CH. www.ecoinvent.org.

4. Gaiddon B & Jedliczka M, *Compared assessment of selected environmental indicators of photovoltaic electricity in OECD cities*. 2006, This technical report has been prepared under the supervision of PVPS Task 10, PVPS Task 10, Activity 4.4, Report IEA-PVPS T10-01:2006, The compilation of this report has been supported by the French Agency for Environment and Energy Management, ADEMEIEA: Hespul, Villeurbanne, France.

5. Hostettler T, *Solarstromstatistik 2005 mit Sonderfaktoren*. Bulletin SEV/AES, 2006. **10**(2006).

6. Classen M, Althaus H-J, Blaser S, Doka G, et al., *Life Cycle Inventories of Metals*. 2007, CD-ROM, ecoinvent report No. 10, v2.0, EMPA Dübendorf, Swiss Centre for Life Cycle Inventories: Dübendorf, CH. Retrieved from www.ecoinvent.org.

7. Frischknecht R, Tuchschmid M, Faist Emmenegger M, Bauer C, et al., *Strommix und Stromnetz*, in *Sachbilanzen von Energiesystemen: Grundlagen für den ökologischen Vergleich von Energiesystemen und den Einbezug von Energiesystemen in Ökobilanzen für die Schweiz*, Dones R, (Editor). 2007, Paul Scherrer Institut Villigen, Swiss Centre for Life Cycle Inventories: Dübendorf, CH. www.ecoinvent.org.

8. Fthenakis V & Alsema E, *Photovoltaics Energy payback times, Greenhouse Gas Emissions and External Costs: 2004-early 2005 Status*. Progress in Photovoltaics: Research and Applications, 2006. **2006**(14): p. 275-280.

9.   Frischknecht R, Hofstetter P, Knoepfel I, Dones R, et al., *Okoinventare für Energiesysteme. Grundlagen für den ökologischen Vergleich von Energiesystemen und den Einbezug von Energiesystemen in Ökobilanzen für die Schweiz.* 1994, Auflage, 1, Gruppe Energie - Stoffe - Umwelt (ESU), Eidgenössische Technische Hochschule Zürich und Sektion Ganzheitliche Systemanalysen, Paul Scherrer Institut Villigen: Bundesamt für Energie (Hrsg.), Bern.

10.   Frischknecht R, Bollens U, Bosshart S, Ciot M, et al., *Ökoinventare von Energiesystemen: Grundlagen für den ökologischen Vergleich von Energiesystemen und den Einbezug von Energiesystemen in Ökobilanzen für die Schweiz.* 1996, Auflage, 3, Gruppe Energie - Stoffe - Umwelt (ESU), Eidgenössische Technische Hochschule Zürich und Sektion Ganzheitliche Systemanalysen, Paul Scherrer Institut, Villigen: Bundesamt für Energie (Hrsg.), Bern, CH. Retrieved from www.energieforschung.ch.

11.   Jungbluth N & Frischknecht R, *Literaturstudie Ökobilanz Photovoltaikstrom und Update der Ökobilanz für das Jahr 2000.* 2000, Programm Aktive Sonnenenergienutzung: Photovoltaik Bericht Nr. 39489, ESU-services for Bundesamt für Energie: Uster. p. 43. Retrieved from www.esu-services.ch.

Mater. Res. Soc. Symp. Proc. Vol. 1041 © 2008 Materials Research Society   1041-R01-05

IR Based Photovoltaic Array Performance Assessment

A. Moropoulou, J. A. Palyvos, M. Karoglou, and V. Panagopoulos
School of Chemical Engineering, National Technical University of Athens, Zografou Campus,
Athens, 15780, Greece

ABSTRACT

In this work infrared thermography was used as a diagnostic tool for the performance assessment of a photovoltaic array integrated on the southern façade of NTUA's Chemical Engineering Building. This grid-connected 50 kWp solar photovoltaic array, installed under an EC Thermie Project (SE-142-97-GR-ES), operates in a standard and hybrid PV-Thermal configuration, meant to save conventional energy. The thermographic system used for the analysis covers the wavelength region 8-12 μm. The thermal images obtained showed that there are notable temperature differences on number PV panels, which may be attributed to material defects, manufacturing faults, module malfunction, and external abuse.

INTRODUCTION

The use of infrared thermography in the analysis of a building's envelope is really developing in recent years. The use of IR data for the calculation of heat transfer coefficients is reviewed in Astarita et al. [1], while Grinzato et al. analyse quantitative infrared thermography in buildings [2]. Avdelidis & Moropoulou discuss the use of IR thermography in buildings of historic interest [3] and Bazilian et al. use thermographic analysis of a building integrated photovoltaic (BIPV) system [4]. This analysis allows for the interpretation of the surface emissivities and operating temperatures, as well as qualitative graphic analysis of temperature gradients. The use of an infrared camera is invaluable in the study of the thermal performance and the relevant parameters in a BIPV system, especially as a diagnostic tool for potential problems, since it involves a relatively quick procedure, which can be accomplished without the need for interrupting system operations.

In this work, which is the result of collaboration between the Laboratory of Materials Science and Engineering and the Solar Engineering Unit of the NTUA's School of Chemical Engineering, infrared thermography is used for the investigation of the BIPV system that has been installed on the southern facade of the Chemical Engineering building of the National Technical University of Athens (NTUA).

THERMOGRAPHIC MEASUREMENTS

The thermographies analysis was performed on the PV arrays of the southern façade of the NTUA's Chemical Engineering building complex (Figure 1). The photovoltaic project was realized under a Thermie Program (SE-142-97-GR-ES) which involved integrating 50 kWp grid-connected solar photovoltaic arrays on the facade and the roof, operating in a standard and

Figure 1: Building façade with the first six PV arrays

hybrid PV-Thermal configuration, meant to save conventional energy (thus limiting pollution normally associated with the latter) and, at the same time, to improve the thermal comfort in the adjacent large laboratory space [5]. The concept of the hybrid operation (Fig. 2) is simple enough : in the winter, the working area air is heated up by rising via free convection in the narrow space between PV's and external wall, and is returned to the interior, thus utilizing the rejected heat due to the low electrical conversion of the PV modules. In the hot season, on the

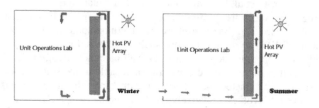

Figure 2: Schematic of the thermal operation of the hybrid system

other hand, the same kind of draft behind the arrays can lead cooler air from the northern side of the laboratory into the laboratory, before venting it into the atmosphere.

The installed arrays, which have a direct-south orientation, use a total of 752 polycrystalline Si modules (Eurosolare PL810), each having a 67 Wp rating, and 8 Fronius IG60 inverters that supply the grid with a maximum power of about 33-34 W.

Such integration demonstrates the potential of solar retrofitting, using appropriate photovoltaic systems and basic heat transfer techniques. As shown above, the technology applied in the hybrid arrays exploits the synergy between the need for cooling the PV cells and the existence of a heating load in the adjacent working space.

The thermographs were taken in September and October of 2006, between the hours 10am to 5pm. For this purpose, a TVS-2000 MKIILW Nippon Avionics infrared camera was used, which operates in the 8-12 μm range of the spectrum.

The thermographs were continuously recorded using a pal video, but also at selected time intervals. The ambient conditions were also recorded (air temperature, relative humidity, velocity).

RESULTS

In the series of images which follows, the left column presents the surface under investigation, while the right column shows the corresponding thermograph. As expected, the best IR thermographic testing images resulted during sunny windless days with low relative humidity.

In the thermograph of Figure 3 hot spots were detected corresponding to areas where the panels have been severely damaged as a result of vandalism, i.e. by stone and bottle throwing.

 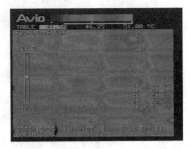

Figure 3: Hot spot detection in a vandalized array

In Figure 4 hot spots were detected in an array without visible external damage on its surface. Thus, the problem must be attributed to internal causes, such as corrosion, or shadow effects, e.g. from the mounting system's aluminium joint covering ribbons. Such hot spots appear in the same areas of the array modules.

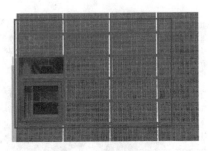 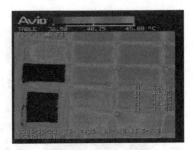

Figure 4: Hot spot detection in an externally undamaged array

Areas around windows exhibiting lower temperatures are shown in Figure 5. Thermal alternation is steep, while the difference between maximum and minimum temperature is substantial. On the upper thermograph two hot spots were also detected, just below the window.

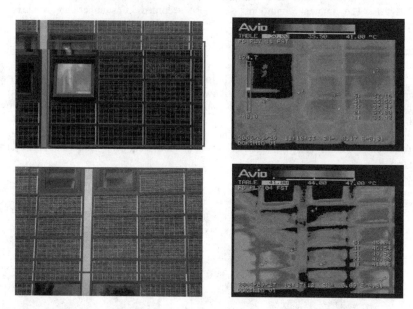

Figure 5: Areas around windows exhibiting lower temperatures

Elongated horizontal hot spots are shown in Figure 6, producing almost vertical temperature gradients in the modules. Main cause of this temperature differentiation is the shadow effect from the horizontal aluminium profiles of the mounting system, under certain angles of solar incidence.

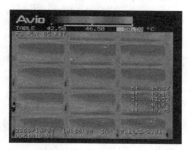

Figure 6: Elongated hot spots in the modules.

DISCUSSION

The identification of a shadow effect caused by elements of the mounting system, was probably the most interesting finding of this thermographic investigation, and convincing proof of the effectiveness of the IR camera as a fault diagnosing tool. The mounting frame turned out to be slightly problematic, because the horizontal metal profiles which cover the joints between modules in the array, under certain solar incidence angles create a thin shadow on the PV material along their upper side. Besides lowering the power output, such shadows cause local overheating and, in the long term, accelerated aging of the modules involved.

The damages which resulted from vandalisms (Fig. 3) caused a substantial decrease in the power output of the particular array, estimated at 13.5 %. Moreover, there is no way of telling how fast the unavoidable corrosion – from water penetration through the cracks – will further degrade the array operation.

CONCLUSIONS

Infrared thermography is a very important diagnostic tool for photovoltaics. Malfunctions, semiconductor material and insulation defects, as well as external damages, can be detected easily and quickly, without complicated procedures. IR thermography can be applied both to large and small-scale systems, in order to help restore or even improve the operation of the array.

REFERENCES

1. T. Astarita, G. Gardone, G.M. Carlomango, C. Meola, A survey on infrared thermography for convective heat transfer measurements, Optics and Laser Technology, 32:593-610, 2000
2. E. Grinzato, V. Vavilov, T, Kauppinen, Quantitive infrared thermography in buildings, energy and buildings, Oxford, Elsevier Science, 1998
3. N.P. Avdelidis, A. Moropoulou, Applications of infrared thermography for the investigation of historic Structures, Journal of Cultural Heritage, 5 (1):119-127, 2004
4. M.D. Bazilian, H. Kamalanathan, D.K. Prasad, Thermographic analysis of a building integrated photovoltaic system, Renewable Energy, 26:449-2\461, 2002
5. THERMIE Project: Integration of Innovative Solar PV-Thermal Systems in the Retrofitting of the NTUA Chemical Engineering Building Complex. (SE/00142/97/GR/ES) http://www.chemeng.ntua.gr/solarlab/THERMIE-en.html

Nanomaterials and
Hydrogen Storage

Mater. Res. Soc. Symp. Proc. Vol. 1041 © 2008 Materials Research Society 1041-R02-01

Nano-Structured Materials to Address Challenges of the Hydrogen Initiative

Vincent Berube[1], and Mildred Dresselhaus[2]
[1]Physics, MIT, 77 Massachusetts Avenue Room 7-008, Cambridge, MA, 02139
[2]Physics and Electrical Engineering, MIT, 77 Massachusetts Avenue Room 13-3005, Cambridge, MA, 02139

ABSTRACT

Since the publication of the 2003 report on Basic Energy Needs for the Hydrogen Economy, many important advances in hydrogen research have occurred, a cadre of enthusiastic researchers has entered the field with great interest shown by students, and private industry has made significant commitment to this technology and investment in its development worldwide. Concurrently, other energy technologies have made major strides forward. These technologies must be evaluated for their scalability, usability, cost and life cycle footprint on the environment. This overview discusses these topics and looks toward the role for the hydrogen economy into our energy future.

INTRODUCTION

Along with food and water, energy availability for the masses is without a doubt a major challenge for the 21st century. Driven by increasing world populations, an even faster increase in the per capita energy demand, a decreasing availability of traditional sources of energy through fossil fuels and the increasing concern about the need to curb the increase of CO_2 into the atmosphere, the need for a transformation to a sustainable energy supply from renewable sources has emerged as a dominant challenge of this century. President Bush in his 2003 State of the Union Message identified this as a major challenge for his administration, as have other national leaders worldwide. As a result of the Bush 2003 State of the Union Message, a hydrogen initiative was subsequently launched by the US Government Funding Agencies.

As a first step, a workshop was held in the spring of 2003, followed by a committee study which resulted in a report [1] which emphasized, firstly the appeal of hydrogen as an energy carrier whose release of energy produces only water as a by product without other pollutants or greenhouse gases, and takes advantage of the high efficiency enabled by hydrogen fuel cells. The report also stressed the challenges for the implementation of the hydrogen economy in terms of the enormous technical challenges to be overcome for its implementation, emphasizing that fundamental breakthroughs would be needed in understanding the physical processes involved in the production, storage and use of hydrogen. Infrastructures to provide for the transportion and distribution of hydrogen would also need to be developed to support a hydrogen economy. Understanding the atomic and molecular processes that occur at the interfaces of materials with hydrogen were identified as crucial to producing the new materials that would be needed for these fundamental breakthroughs to occur. The report goes on to say that the discovery of the new materials, new chemical processes and new synthesis techniques that would be required could only be achieved by initiating a major basic research program with these objectives.

Such a research program was subsequently launched by the Basic Energy Sciences Office of the US Department of Energy (DOE) following the recommendations of the report, working in close collaboration with the Energy Efficiency and Renewable Energy Office of the DOE, thereby uniting the basic and applied science thrusts through a highly interdisciplinary effort involving chemistry, physics, biology and engineering, all working together to solve the multitude of challenges and opportunities identified in the report. From these efforts, major research advances have occurred over a short period of time in the areas of hydrogen production, storage, and fuel cell performance, amplified by the corresponding efforts occurring worldwide. The enthusiastic response of the research community and the great interest of students in joining this effort has been noteworthy, leading to a series of other workshops, studies and initiatives in other areas of energy research and development. These efforts have also led to significant scientific and technological advances. Concurrently, industry has launched major initiatives so that the playing field is rapidly changing as breakthroughs are occurring throughout the energy science and technology landscape. Developments in biofuels, solar energy, battery technology, hybrid technology and clean coal, to name but a few, are rapidly changing the energy outlook into the future. Even if a combination of technologies is likely to be necessary to address our future energy needs, present trends show that those technologies which offer immediate market value are most likely to attract additional investments and to get developed even faster. In the present brief report, emphasis is given to an attempt to identify an evolving role for the hydrogen economy within the larger energy challenge and what types of breakthroughs are needed to succeed.

STRATEGIC ISSUES

Based on the DOE hydrogen requirements for the years 2010 and 2015 (Table 1), the technology gaps for hydrogen as an energy carrier were identified (Fig. 1) and research directions for bridging these technology gaps were suggested in the Basic Research Needs Report [1]. In the meantime, the auto industry worldwide has taken a hydrogen-based vehicle seriously and has moved rapidly in getting hydrogen fuel cell automobiles on the highways to gain experience with this new technology, using presently available methods for hydrogen production and storage. Industry has focused mainly on accelerating hydrogen fuel cell development and the infrastructure needed for carrying out a hydrogen vehicle test program. While methods for hydrogen production from natural gas are presently adequate for automotive needs, the use of a fossil fuel natural gas precursor defeats the long term goal of using a sustainable, renewable energy source to provide the large increase in hydrogen production (20-fold by the estimate in Figure 1) that would be required for transportation use. The development of a renewable route for large scale hydrogen production by methods, such as splitting water in a closed cycle water-hydrogen process or by a biologically-inspired process remains a long term challenge where there are presently large opportunities for the research community.

The on-board storage of hydrogen to match US consumer appetites for a 500 km (~300mi) range for their family vehicle had been identified as the greatest challenge to the implementation of a hydrogen economy, because even the filling of the present fuel tanks of an automobile with liquid (or solid) hydrogen filling would fall short of meeting the DOE 2015 targets (see Table 1). H_2 molecules strongly repel each other because of their spherical electronic shell and have a very low mass density even in the liquid or solid phase. This is why storage materials that bind hydrogen usually have a higher hydrogen density than liquid hydrogen since the bond between hydrogen and the storage material changes the H_2 electronic structure and allows H_2 molecules or atoms to come closer than they would in a gaseous or liquid phase.

The auto industry has taken a different approach toward addressing consumer appetites and is using increased operating efficiency and hybrid vehicle technology to manage the storage requirements. Using this approach, Toyota has recently demonstrated by a run from Osaka to Tokyo a 550 km (350mi) range for its hydrogen fuel cell vehicles based on presently available compressed hydrogen gas cylinder technology. Although researchers from the auto industry are anxious for the academic community and government supported research laboratories to come up with a chemisorbed or physisorbed hydride solution for hydrogen storage, the auto industry does not now see the hydrogen storage problem as a technical show-stopper, though widespread public acceptance of the hydrogen gas cylinder technology has not been seriously tested. Thus, the auto industry is looking to the research community for major breakthroughs in renewable hydrogen production, reversible solid state hydrogen storage and higher efficiency hydrogen fuel cells to help make widespread adoption of the hydrogen fuel cell vehicle option a reality by mid-century. The arguments on the central role that new materials, and especially nanostructured materials, will play in these breakthroughs, as presented in the 2003 hydrogen report [1], remain valid through the present time.

What has changed in the interim is the vital role that industry is now playing and the need for the research community to be in close contact with industrial R&D developments, and to play a role in the incubation of start-up companies to develop the new technology options that will be provided by future suppliers to the auto companies. Thus, one strategic issue for the planning of hydrogen research is the coordination, not only between basic and applied research by the multidisciplinary players, but also to look for opportunities where academic and national laboratory research could have a large impact on future industrial product development. An effective way to accomplish those goals and to raise awareness for the hydrogen economy could be to develop niche markets for current hydrogen technologies. For examples, there could be opportunities in portable energy storage where hydrogen technology could currently compete on both a cost and storage capacity basis with current battery technologies. The knowledge and capital generated from those niche markets successes could further fuel the R&D necessary to develop new materials that could satisfy the evolving requirements of the auto industry. We have seen the impact that private investment can have on the development of a technology in other fields such as solar energy, where competition for market share has driven research towards roadmaps that follow the same pattern as Moore's law has set in motion for semiconductor electronics.

A second strategic issue concerns scale. Projections of global energy needs imply a doubling in overall energy demand and a tripling of the electricity demand by the year 2050 relative to the year 2000. The only renewable energy sources with sufficient capacity to meet these growing energy demands is solar energy. An increase from the present 14TW to 28-30TW by 2050 is expected to come from solar energy used for generating electricity (photovoltaic), providing fuels (biofuels, water splitting, close cycle synfuels), and supplying space and water heating (solar thermal). In this big picture, with solar electric, solar fuel and solar thermal as the energy sources, electricity and hydrogen are identified as complementary energy carriers. When thinking of hydrogen as a chemical carrier of energy, its role in energy storage from the electric grid emerges as an interesting opportunity, as does the generation of close-cycle renewable synfuels using hydrogen from H_2O and carbon from CO_2 to produce a hydrocarbon fuel using sunlight as an energy source.[3] The latter research direction, denoted by "transformation and recycling of CO_2 into a new material", was identified in the Declaration issued by the First World Materials Summit held in Lisbon in 2007.[4]

Table 1: Requirements for a hydrogen fuel cell automobile. Source Milliken (2003).

Targeted Factor	2005	2010	2015
Specific energy (MJ kg)	5.4	7.2	10.8
Hydrogen (wt%)	4.5	6.0	9.0
Energy density (MJ/L)	4.3	5.4	9.72
System cost ($/kg/system)	9	6	3
Operating temperature (°C)	-20/50	-20/50	-20/50
Cycle life-time (absorption/desorption cycles]	500	1,000	1,500
Flow rate (g/s)	3	4	5
Delivery pressure (bar)	2.5	2.5	2.5
Transient response (s)	0.5	0.5	0.5
Refueling rate (kp/H2/min)	0.5	1.5	2.0

The need for breakthroughs with high impact follows from the huge scale of the energy challenge involving a multi-trillion dollar business worldwide. Therefore major emphasis must be given to those research directions which will have the potential for large orders of magnitude impacts.

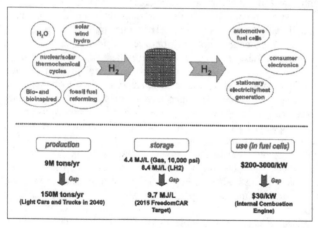

Figure 1: The technology gaps in hydrogen production, storage, and end use in a hydrogen economy.[2]

This brings to mind Moore's law which has provided road-maps for the electronics, optoelectronics and magnetic information storage industries for several decades. To have comparable impact on the energy industry, a Moore's law road-map for the Energy Industry is needed. Here new materials will play a vital role, especially nanomaterials, because of the greater ability to modify and control their properties by varying the material's size and composition, their greater surface area to promote catalysis which is based on an exponential *exp(−E/kT)* dependence, and the independent control of materials parameters which are interdependent in 3D systems.

(a)   (b)

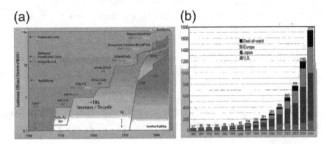

Figure 2: Examples of energy industries showing aspects of Moore's law behavior: (a) solid state lighting efficiency, (b) photovoltaic cell production in MW.

In fact, Moore's law has started to infiltrate the energy industry. One example is solid state lighting where the [lumens/watt] emission from light emitting diodes has followed a Moore's law path in the last 30 years [Figure 2(a)]. This technology now requires half the electrical energy of an equivalent incandescent lamp for a given light output. Solid state lighting is thus expected to have a major impact on the drive toward improving energy efficiency, since residential and industrial lighting currently accounts for 22% of electricity use in the US. Research is actively occurring to improve light quality, to lower cost and to find new uses for this transformational technology, uses that are different relative to the technology it replaces. A second example of Moore's law is photovoltaic (PV) cell production [Figure 2(b)] which has had a worldwide annual growth rate of ~30% for the last decade, but in which the USA has not been a major player. Recent advances in photovoltaic technology, using three junction devices which capture the solar spectrum very well, have achieved over 40% efficiency in PV conversion [5].

This technology, using a solar concentration of 240 suns, requires only 10^{-3} of the "real estate" of conventional solar cells, is well positioned for both scale-up and new applications areas for photovoltaics. Even though the technology is quite complex and requires many semiconducting layers, Spectrolab (a subsidiary of Boeing Corp) has been developing these technologies for commercialization. They have recently released a road-map for the scale-up of the production for 2010. Their device has 40% efficiency at a cost of less that $0.15/kWh and increased performance and lower cost are projected for the future. This compares well with current technology. In 2007 the lowest PV module production cost for PV technology was First Solar's CdTe PV at $1.1/Wpeak and the lowest system installed cost was the 40 MW JuWI system in Saxony, GR. [23]. This cost corresponds to $0.16/kWh [24]. The Spectrolab solar concentration technology could be used for both power generation in power plants or on the rooftops of residential homes, with a potential for major future impact on electricity production and energy efficiency. Since sunlight is intermittent, there could be interesting opportunities for hydrogen as an energy storage agent to be used in conjunction with this technology.

Another interesting direction where large-scale impacts on energy are occurring is in thermoelectric conversion where increases in the thermoelectric figure of merit and scale-up to samples with higher thermal capacities have been demonstrated. As a result, industrial development in this field is booming with about one million cooling/heating thermoelectric seats sold in 2007 for automotive use. When these seats are used in hybrid cars where fuel efficiency is readily monitored, it has been found that the local cooling of passengers by the thermoelectrically equipped seats causes a major decrease in the need for air conditioning for passenger comfort, resulting in a payback of less than 1 year for the thermoelectric car seats, with subsequent cost savings in future fuel consumption.[6] It would be interesting to see what the effect of thermoelectric car seats would be on the efficiency of hydrogen fuel cell autos.

The device utilization of the discovery of highly efficient carrier multiplication in semiconductor nanocrystals [7] allowing as many as 6 electron-hole pairs to be produced by a single optical photon incident on a PbSe nanocrystal is now being explored and may eventually result in enhanced photovoltaic device efficiency. If this scientific advance results in improved photovoltaic device efficiency, this may open new opportunities for hydrogen as an energy storage agent.

Finally, high throughput combinatorial screening allows a route for both experiment and theory to scan many variants of multi-component materials by composition, to optimize a material for a given property while at the same time allowing rapid measurement of several other properties of the material in the compositional range where the desired property is optimized. Such capabilities are necessary since a number of properties of a material affect its ultimate device performance, and these properties therefore need to be jointly optimized. For example, a material, which has excellent thermoelectric performance but is toxic, would not make it in the marketplace. A good example of high throughput screening is the use of optical properties of metal hydrides to screen for potential alloy candidates that offer good hydrogen storage properties. Upon hydriding, many hydrides exhibit a reflectivity change when they make a transition from a conducting to an insulating material. The reduction of the reflectivity can thus be used to identify promising materials for hydrogen storage and to then study their properties, such as their kinetics and sorption/desorption temperature, properties that are also required for hydrogen storage materials. The reduction of their reflectivity can also be used at different temperature and pressures to determine the enthalpy of formation of hundreds of hydrides at the same time. This is essential in determining which materials offer the required properties for automotive applications. [21]

STRATEGIES FOR A HYDROGEN ECONOMY

With the principles outlined above in mind, we can identify a number of breakthroughs that have the potential for high impact on the hydrogen economy. As mentioned above, the use of improved catalysts has the potential for high impact because the probability of a reaction to take place has an $exp(-E/kT)$ dependence where E is the activation energy (energy barrier) that can be significantly reduced by a catalyst. Thus, a promising strategy is the search for new catalysts that lower the energy barriers for chemical reactions, that can be made in the optimal range of small sizes (usually in the 2–5nm range), and can be composed of cheaper and more plentiful elements. An example where such a specially tailored catalyst has been developed for the hydrogen economy is the Pt_3M catalyst. Density functional theory was first used to establish the concept of using a Pt surface layer of the catalytic particle to rapidly dissociate a hydrogen molecule. The introduction of a first subsurface layer with a PtM composition then provides a mechanism for attaching atomic hydrogen more easily.[8] Such an approach can provide both strong binding and also rapid release of hydrogen when used either in physisorption or chemisorption-based storage applications. Variants of this concept could have an impact on hydrogen production, storage, and use in fuel cells. An implementation of this general concept has recently been made to increasing the catalytic activity of Pt by a factor of 10 in the oxygen reduction reaction by using a surface Pt layer and a subsurface PtNi layer to break the O–O bonds to form O–H bonds. Weak surface bonds prevent the splitting of O–O bonds, while strong surface bonds attract guest species to adhere to the surface, thereby blocking access of other reactants to the catalyst. In the case of the oxygen reduction reaction, the 10-fold increase in catalytic activity for the oxygen reduction reaction which occurs at the anode of hydrogen fuel cells was achieved by using both the (111) crystal orientation of the catalytic particle and its compositional variation. [9] Research on catalysts and catalytic properties of certain materials or structures will be one of the most important fields to investigate for a successful hydrogen economy. For hydrogen production, purification and utilization in a Fuel Cell, the catalysts used are often expensive materials, such as Pd, Ag and Pt that make the systems expensive and would place limitations on the scalability of certain technologies. It is thus important to investigate potential substitutes for those materials. For example, hydrogen purification can be achieved by using a Pd-Ag membrane that is permeable to H_2 but not permeable to contaminants such as water or CO_2. To replace Pd and Ag, research has been done on common metallic alloys made from Ni, V and Ti that could be used to purify H_2 at a much lower cost. [20]

A number of other impressive advances in hydrogen research have been made in the laboratory at the research level, and a small number are cited here as examples. One noteworthy example is the identification of a route to increase the tolerance of hydrogen production by a genetically modified Fe–Fe hydrogenase bacterial structure that yields a 100-fold increase in H_2 activity relative to the natural algol enzyme. Simplified and robust analogs of bacterial hydrogenase have the potential to lead to the development of a commercial-scale hydrogen production route that may be scalable to large scale, self-sustaining and cost effective production.[10] Genetic modification in algae to reduce the photosynthesis/respiration capacity ratio can also lead to the direct production of hydrogen from plants by optimizing the solar conversion efficiency and reducing energy waste. Micro algae have the advantage of being self repairing/reproducing and cheap. Such a technology could also be installed in a CO_2 waste stream of a conventional coal plant as a way to reduce the CO_2 emissions of those plants. This is a great example of synergistic behavior between two technologies that improve the total performance of the system [17].

Most of the research on hydrogen storage is still being done either on chemisorption or physisorption processes. There have been many interesting developments to improve the kinetics and to reduce the release temperature of chemisorption-based hydrogen storage materials and to increase the stability and bond strength of physisorption-based hydrogen storage materials. An interesting approach to lowering the release temperature of hydrogen in a metal hydride is through increased destabilization by using a second compound in a chemical reaction. For example $LiNH_2 + LiH \longrightarrow Li_2NH + H_2$ releases hydrogen at ~150°C which is significantly lower than $LiNH_2$ (at 200°C) or LiH (at 500°C). This study is of significant scientific interest. However the storage capacity for the joint reaction is only 6.5%, which could be too low for commercial development [12, 13] On the other hand, the destabilization pair of $LBH_4 + MgH_2$ with a storage capacity of 11.5% could be more interesting for further commercial development.[14] It is also possible to improve the hydrogen storage properties of certain materials and to simultaneously increase the stability of the nanostructures by using scaffolding methods. Insertion of LiBH$_4$ in a carbon aerogel scaffold, for example, leads to kinetics that are 50 times faster than desorption from a carbon substrate. The advantage of the nanoscaffold is to limit the size of the hydride material particles and to prevent their agglomeration while at the same time offering the porosity needed for good diffusion of the hydrogen into or out of the storage materials. The same scaffolding method used for MgH$_2$ in a carbon aerogel, previously wetted with Ni, allowed for H$_2$ desorption at 275°C compared to more than 400°C for bulk magnesium. [18]

Some interesting improvements in physisorption-based materials have also been made. One of the fastest moving fields is in metal organic frameworks (MOFs). MOFs are crystalline compounds consisting of metal ions or clusters coordinated to rigid organic molecules to form one-, two-, or three-dimensional structures that are extremely porous and show unusually large surface areas (5000m^2/g has been reported [19]). The main advantage of MOFs remains their large porosity and surface area that allows for fast kinetics and good storage properties. Recently MOF 177 [19] was reported to have a 7.5 wt% of H$_2$ at 77 K and 70 bar. Hundreds of those MOFs have already been discovered or predicted by first principles simulations, and new types of complementary structures that could have even better binding properties are now being investigated (Zeolite Imidazolate Frameworks and covalent Organic Frameworks). In all cases, a better understanding of the binding mechanism will be necessary to increase the bond strength between H$_2$ and the frameworks that might allow the room temperature storage requested by the DOE.

Some new ideas have recently been introduced with an aim toward increasing the temperature of operation of PEM membranes and increasing the power density of the fuel cell operation. Some membranes have been developed that conduct protons at temperatures up to 200°C in the absence of water [15]. A new class of chemically cross-linked membranes, fabricated at low temperature from liquid precursors, significantly enhance proton conductivity by allowing additional acid loading, enhance thermal and mechanical stability by increased cross-linking, while at the same time increasing electrical and chemical exchange with the electrode by enhancing the effective surface area.[16]

Many of the recent advances in hydrogen research are still at an early stage of developement with further progress in understanding and in materials performance expected in the near term. Applications to industrial products are expected to follow. As the technologies needed for a large scale usage of hydrogen are developed, measures should be taken to promote the acceptance of a hydrogen economy by the general public. Recent studies amongst students, state officials, the general public, and the potential end users of hydrogen show that many misconceptions about the safety or about the physical properties of hydrogen are still common. Those studies also reveal that there exists an inverse correlation between the will of people to accept hydrogen as an energy carrier and the level of their knowledge about hydrogen [22]. As we transition to a hydrogen economy or as hydrogen starts penetrating small niche markets, it will thus be important to put in place programs to educate the different segments of the populations that will be exposed to hydrogen. A joint effort between private companies and the government is most likely to be the best approach to reaching such a goal.

CONCLUDING REMARKS

Because of its special and unique attributes, hydrogen is likely to be one of a mix of future sustainable energy technologies. New materials and nanoscience are necessary for the development of the potential of this technology as they are for the future development of many of the other pertinent energy technologies. The strong interplay between basic and applied sciences, interdisciplinary approaches and the coupling between theory and experiment are all vital. Working closely with industry will be important for identifying research directions with high potential impact. Finding niche markets for early hydrogen technology and bringing those technologies to market will further increase R&D investment from the private sector who will see the possibility for profit. The development of those niche markets would also help develop the infrastructures needed for a hydrogen economy and allow some of the cost to be carried by private investors. Attention to major advances in other key technologies is equally important for the identification of new priority directions for hydrogen R&D. Because of the highly complementary focus of energy research in different countries, based on their different climatic and cultural constraints, international cooperation and networking should be encouraged and supported. Linking to and coordinating between international groups promoting materials research for energy applications regionally and internationally would be important, so that policy makers worldwide get a clear message about progress in hydrogen research and its potential contribution to the larger picture of providing a sustainable energy supply world-wide.

Developing sustainable energy technologies that can have a major impact and implementing them will require much effort. The Lisbon declaration of the World Materials Summit [4] highlights key steps that must be addressed to ensure a smooth and successful transition to new energy technologies. It highlights the importance of

1. Strategic planning (roadmaps) in the development of new and improved materials and the products for future energy technologies.
2. Bringing together leading academic, public sector and industrial scientists to discuss important technical issues and to ensure that key problems are tackled in a swift and effective manner.
3. Identifying and training a new generation of young international leaders.
4. Promoting major new international collaborative materials research programs relevant to future energy technologies.
5. Providing information to global, regional and national policy makers, and to investment analysts in the energy sector.
6. Ensuring that manufacturers in the energy sector, especially small and medium enterprises, have the best possible access to information related to innovative materials developments.
7. Interfacing with other key international organizations relevant to the energy sector.
8. Stimulating public interest in, and awareness of, energy-related issues.
9. Attracting and nurturing the young generation of scientists and engineers to meet the mega challenge of clean energy sustainability and growth through providing a clear picture of the challenges, opportunities and career paths for energy-related research.

Implementing major changes in our energy consumption habits will demand much more than breakthrough technologies. It is a challenge that will take time and the systematic commitment over time of the academic, private, and public sectors. This is why communication between the different sectors is important for sharing specific needs and knowledge which will ultimately accelerate technology development and implementation.

ACKNOWLEDGMENTS

The authors acknowledge G. Dresselhaus and G. Crabtree for valuable discussions. The authors acknowledge support under DE-FG02-05ERR46241.

REFERENCES

1. G. W. Crabtree, M. S. Dresselhaus, and M. V. Buchanan, Basic Research Needs For the Hydrogen Economy (Office of Basic Energy Sciences, Department of Energy, BES, Washington DC, 2003).
2. G. W. Crabtree, M. S. Dresselhaus, and M. V. Buchanan, *Physics Today* 57(12), 39–44 (2004). December.
3. Koji Hashimoto, N. Kumagai, K. Izumiya, Z. Kato, Materials and technology for global carbon dioxide recycling for supply of renewable energy and prevention of global warming, 2007.
4. Declaration issued by the First World Materials Summit, Lisbon, Portugal 2007 http://www.spmateriais.pt/LISBON%202007%20DECLARATION.pdf.
5. R. R. King et al, Appl. Phys. Lett. 90, 183516 (2007).
6. Lou Bell report at the Industrial Physics Forum, Seattle, WA, Oct 2007.
7. R.D. Schaller and V. I. Klimov, *Phys. Rev. Lett.* 92, 186601, (2004).
8. J. Greeley and M. Mavrikakis, Alloy catalysts designed from first principles, *Nature Materials*, 3, 810 (2004).
9. V. R. Stamenkovic et al, *Science* 315, 497 (2007).
10. P. W. King et al, *Proc. SPIE* vol 6340. 63400Y (2006).
11. G. W. Crabtree and M. S. Dresselhaus, *MRS Bulletin*: Energy Issue page in press (2007).
12. P. Chen, Z. Xiong, J. Luo, J. Lin, K.L. Tan, *J. Phys. Chem. B* 107, 10967 (2003).
13. J.F. Herbst, L.G. Hector, Jr., *Phys. Rev. B* 72, 125120 (2005).
14. J.J. Vajo, G.L. Olson, *Scripta Mater.* 56, 829 (2007).
15. J.A. Asensio, S. Borrs, P. Gmez-Romeroa, *Electrochim. Acta* 49, 4461 (2004).
16. Z. Zhou, R.N. Dominey, J.P. Rolland, B.W. Maynor, A.A. Pandya, J.M. DeSimone, *J. Am. Chem. Soc.* 128, 12963 (2006).
17. A. Melis, Material Issues in Photobiological Hydrogen Production, *Proceedings INTERNATIONAL SYMPOSIUM ON MATERIALS ISSUES IN A HYDROGEN ECONOMY* (2007).
18. J. J. Vajo, A. F. Gross, R. D. Stephens, T. T. Salguero, S.. L. Van Atta , P. Liu, Hydride Chemistry in Nanoporous Scaffolds, *Proceedings INTERNATIONAL SYMPOSIUM ON MATERIALS ISSUES IN A HYDROGEN ECONOMY* (2007).
19. A. G. Wong-Foy, A. J. Matzger, O. M. Yaghi, J. Am. Chem. Soc. 128, 3494 (2006).
20. T. Adams, P. S. Korinko, Alternative Materials to Pd Membranes for Hydrogen Purification, *Proceedings INTERNATIONAL SYMPOSIUM ON MATERIALS ISSUES IN A HYDROGEN ECONOMY* (2007).

21. B. Dam, Hydrogenography: A combinatorial thin film approach to identify thethermodynamic properties metal hydrides, *Proceedings INTERNATIONAL SYMPOSIUM ON MATERIALS ISSUES IN A HYDROGEN ECONOMY* (2007).
22. "National   Hydrogen   Energy   Roadmap" http://www1.eere.energy.gov/hydrogenandfuelcells/pdfs/national_h2_roadmap.pdf. (visited on 11/20/07).
23. Fairley, *MIT Tech Rev.*, July 27, (2007).
24. Zweibel at al., *Scientific American*, 64-73 (Jan. 2008).

Mater. Res. Soc. Symp. Proc. Vol. 1041 © 2008 Materials Research Society   1041-R02-02

High-Surface-Area Biocarbon for Reversible On-Board Storage of Natural Gas and Hydrogen

Peter Pfeifer[1], Jacob W. Burress[1], Mikael B. Wood[1], Cintia M. Lapilli[1], Sarah A. Barker[1], Jeffrey S. Pobst[1], Raina J. Cepel[1], Carlos Wexler[1], Parag S. Shah[2], Michael J. Gordon[2], Galen J. Suppes[2], S. Philip Buckley[3], Darren J. Radke[3], Jan Ilavsky[4], Anne C. Dillon[5], Philip A. Parilla[5], Michael Benham[6], and Michael W. Roth[7]

[1]Department of Physics, University of Missouri, Columbia, MO, 65211

[2]Department of Chemical Engineering, University of Missouri, Columbia, MO, 65211

[3]Midwest Research Institute, 425 Volker Blvd., Kansas City, MO, 64110

[4]Advanced Photon Source, Argonne National Laboratory, 9700 S. Cass Ave., Argonne, IL, 60439

[5]National Renewable Energy Laboratory, 1617 Cole Blvd., Golden, CO, 80401

[6]Hiden Isochema Ltd., 231 Europa Blvd., Warrington, WA5 7TN, United Kingdom

[7]Department of Physics, University of Northern Iowa, Cedar Falls, IA, 50614

ABSTRACT

An overview is given of the development of advanced nanoporous carbons as storage materials for natural gas (methane) and molecular hydrogen in on-board fuel tanks for next-generation clean automobiles. The carbons are produced in a multi-step process from corncob, have surface areas of up to 3500 m^2/g, porosities of up to 0.8, and reversibly store, by physisorption, record amounts of methane and hydrogen. Current best gravimetric and volumetric storage capacities are: 250 g CH_4/kg carbon and 130 g CH_4/liter carbon (199 V/V) at 35 bar and 293 K; and 80 g H_2/kg carbon and 47 g H_2/liter carbon at 47 bar and 77 K. This is the first time the DOE methane storage target of 180 V/V at 35 bar and ambient temperature has been reached and exceeded. The hydrogen values compare favorably with the 2010 DOE targets for hydrogen, excluding cryogenic components. A prototype adsorbed natural gas (ANG) tank, loaded with carbon monoliths produced accordingly and currently undergoing a road test in Kansas City, is described. A preliminary analysis of the surface and pore structure is given that may shed light on the mechanisms leading to the extraordinary storage capacities of these materials. The analysis includes pore-size distributions from nitrogen adsorption isotherms; spatial organization of pores across the entire solid from small-angle x-ray scattering (SAXS); pore entrances from scanning electron microscopy (SEM) and transmission electron microscopy (TEM); H_2 binding energies from temperature-programmed desorption (TPD); and analysis of surface defects from Raman spectra. For future materials, expected to have higher H_2 binding energies via appropriate surface functionalization, preliminary projections of H_2 storage capacities based on molecular dynamics simulations of adsorption of H_2 on graphite, are reported.

INTRODUCTION

According to the State Alternative Fuels Plan [1] of the California Air Resources Board and California Energy Commission, adopted October 31, 2007, in response to Assembly Bill 1007, the State of California will take action to increase its use of natural gas (NG, methane, CH_4) as motor fuel from currently 0.6% to 19% (aggressive scenario) of the state's on-road transportation fuels by 2050. According to the 2006 Hydrogen Posture Plan [2] of the U.S. Department of Energy (DOE) and U.S. Department of Transportation, a projected 37% of light-duty vehicles in the U.S. will be hydrogen fuel cell vehicles by 2050.

Both fuels meet the "No Net Material Increase in Emissions" standard [1,3]. For both fuels, the "holy-grail" on-board tank [4] is a lightweight, flat-panel (conformable) tank, under the floor or in other unused space of an automobile, that has a driving range of more than 300 miles, can be fueled in less than 3 minutes, and requires a minimum of auxiliary on/off-board equipment and infrastructure. However, most current natural-gas vehicles run on compressed natural gas (CNG), with on-board NG stored in bulky, heavy-walled cylinders at 250 bar (3600 psig). Likewise, hydrogen fuel cell vehicles under development store hydrogen in cylinders at 350-700 bar. Such high-pressure tanks are difficult to integrate within the space available in a passenger car and give the vehicle a limited fuel storage capacity, whence limited driving range. This makes on-board storage a major barrier to the use of NG and hydrogen for advanced transportation, in the transition to non-petroleum transportation fuels.

The key to remove this barrier is to store the fuel in a porous solid, as adsorbed natural gas (ANG) or adsorbed hydrogen, designed to hold the fuel at low pressure (e.g., 35 bar, 500 psig) with a storage density comparable to that in a high-pressure tank. The low pressure allows for thin tank walls and a conformable shape (Figure 1). In this paper we report significant progress toward this goal, based on novel nanoporous carbons developed by our team (Alliance for Collaborative Research in Alternative Fuel Technology—ALL-CRAFT [5]), both for CH_4 and H_2. Earlier work found nanoporous carbons crisscrossed by a nearly space-filling network of channels, ~1.5 nm wide [6], close to the optimal width of 1.1 nm for CH_4 storage [4].

For CH_4, the DOE in 2000 defined the volumetric storage target at 35 bar and room temperature as 180 V/V, that is, the adsorbent stores 180 its own volume of methane at 1 bar and 298 K [7]. This translates into 118 g CH_4/liter carbon. Characteristics of earlier ANG projects, prior to the ALL-CRAFT materials, are summarized in Table I.

For H_2, the 2010 DOE volumetric and gravimetric targets (system targets) are 45 g H_2/liter tank and 60 g H_2/kg tank ("6.0 mass%") at 293 K.

EXPERIMENTAL DETAILS

Nanoporous carbons were made by pyrolysis of ground waste corncob using a proprietary multi-step procedure [8]. Samples were prepared in granular form. For the prototype tank (Fig-

Figure 1. Left: Schematic of a multi-cell, conformable ANG tank, filled with carbon briquettes. Center and right: ALL-CRAFT prototype ANG tank and fuel delivery system, currently being road-tested.

Table I. Comparative characteristics of ANG projects up to 2006 [9]. Not included in the table is methane storage work done by Yaghi's group on metal-organic frameworks [10]. The storage capacities reported in Figures 2-3 below correspond to the entry "Tank uptake V/V" in the table.

PARAMETERS AND CONDITIONS	AGLARG (Atlanta Gas Light Adsorbent Research Group)	EU FP5 LEVINGS program (coordination by FIAT)	OAK RIDGE NATION. LABO-RATORY (ORNL)	HONDA MOTORS	UNIVERSITY OF PETROLEUM CHINA (UPC)		Brazilian Gas Technology Center (CTGÁS)
Years	1990-1999	1997-2000	? -2000	From 2000	1994-95		From 2000
Investigation method	Chrysler B-van, Dodge Dakota Truck	FIAT Marea, On-board, field testing	Laboratory Investiga-tions	Tank development Adsorbent - laboratory tests	Car XIALI 713IU On-board, field testing		Laboratory investigation on full-size prototype
Pressure, bar	35-40	35-40	35	35	50	125	35-40
Tank uptake V/V	150 in laboratory condition, 142 on-board	123	150	155	100-110	170-180	130-150
Tank delivery V/V (to engine)	135 (approx)	107	Not relevant	-	Un-known	Un-known	Unknown
Adsorbent presumed cost	Prohibitive	High, but about 10 times less than the AGLARG	Supposedly very high	Supposedly similar to AGLARG	Un-known	Un-known	Unknown
Vessel (tank) design features	Multicell of extruded aluminum	Multicell of steel tubes	Small laboratory vessel of volume 0,05 L.	Multicell	Un-known	Un-known	Cylindrical form with volume 30 liters

ure 1 b,c), the granular carbon was pressed into disk-shaped briquettes (monoliths), 3.5 inches in diameter and 0.5-1.0 inches thick, using a binder and heat treatment. We manufactured over 300 briquettes (~25 kg), loaded them in the prototype tank and fuel delivery system constructed by the Midwest Research Institute, and installed the system on a NG vehicle (Ford F-150 bifuel pickup truck) on loan from the Kansas City Office of Environmental Quality. The tank was showcased in Kansas City, 2/16/07 [11], and has been on the road in Kansas City ever since.

Methane uptakes were measured as excess adsorption, m_{ads}^e (total mass of methane in the pore space minus the mass of bulk methane that would be present in the absence of adsorption) on a custom-built gravimetric instrument (sample masses 1-5 g) and on a custom-built volumet-ric instrument for briquettes (Sievert apparatus). Excess adsorption was converted into total amount stored, m_{st} (mass of adsorbed and nonadsorbed methane in the pore space), using

$$m_{st} = m_{ads}^e + (\rho_a^{-1} - \rho_s^{-1})\rho_{gas}m_s \tag{1}$$

where ρ_a, ρ_s, ρ_{gas}, and m_s are the apparent density of the sample (including pore space), skeletal density of the sample (without pore space), density of bulk gas, and mass of the sample, respec-tively. Hydrogen uptakes were measured as excess adsorption on the custom-built gravimetric instrument (University of Missouri), a Hiden IGA-001 instrument (Hiden Isochema Ltd.), and a Hy-Energy PCTPro2000 instrument (National Renewable Energy Laboratory), and converted into total amount stored using Eq. (1), just as in the methane case.

Surface areas and pore-size distributions were obtained from N_2 adsorption at 77 K on an Autosorb-1-C instrument (Quantachrome), computed from density functional theory (DFT) analysis of the N_2 isotherm (Quantachrome). Scanning electron microscopy (SEM) was performed on a Hitachi S-4700 FESEM instrument, with beam energy set to 5 kV and a small working distance (3-4 mm). Transmission electron microscopy (TEM) was performed on a JEOL 1200EX TEM instrument, with beam energy 100-120 kV. Small-angle x-ray scattering data was collected on Beamline 32-ID-B USAXS (Ultra-Small-Angle X-Ray Scattering) at the Advanced Photon Source, Argonne National Laboratory.

Temperature-programmed desorption (TPD) spectra for hydrogen and Raman spectra were measured at the National Renewable Energy Laboratory. In the TPD experiments, a sample was sequentially degassed in vacuum with steps up to 523 K, in order to ensure that no adsorption sites were blocked. After each degassing step, the sample was exposed to 500 torr H_2 at room temperature, cooled to ~190 K, followed by evacuation of the chamber. H_2 evolution was then monitored with a mass spectrometer as the sample was heated with a linear heating rate.

RESULTS AND DISCUSSION

Methane Storage

Figures 2 and 3 show methane storage isotherms (total amount stored) for a typical ALL-CRAFT carbon briquette, and volumetric vs. gravimetric storage isotherms of our best samples to date (S-33/k and Batch 5.32). The data shows that the carbon-filled tank stores 5-6 times more methane than a tank without carbon at 35 bar, despite the fact that the carbon skeleton occupies 20-30% of the tank volume. To store 118 g CH_4/liter without adsorbent, the tank pressure would have to be 180 bar, much more than what a flat tank can bear. The target pressure of 35 bar (500 psig) of the ANG tank equals the pressure in typical NG pipelines and, therefore, elimi-

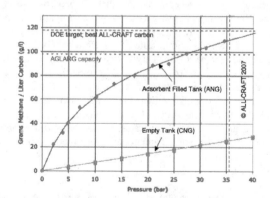

Figure 2. Low-pressure volumetric CH_4 storage isotherm for a typical ALL-CRAFT carbon briquette at 293 K, compared to previous work (AGLARG, see Table I), the DOE target of 118 g CH_4/liter adsorbent [7], and a tank without adsorbent.

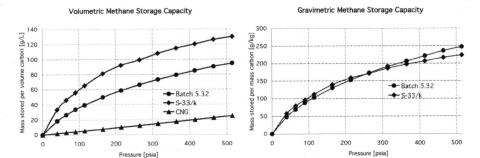

Figure 3. Left: Best volumetric storage capacity, 130 g CH₄/liter carbon (199 V/V) at 35 bar and 293 K, for Sample S-33/k. Right: Best gravimetric storage capacity, 247 g CH₄/kg carbon at 35 bar and 293 K, for Sample Batch 5.32. The reversal of maximum storage capacity, if one goes from volumetric to gravimetric capacity, Eqs. (2) and (1), is due to the difference in apparent density of the two samples: $\rho_a = 0.58$ g/cm³ for S-33/k and $\rho_a = 0.38$ g/cm³ for Batch 5.32. The respective porosities are 0.71 and 0.81.

nates costly compression of NG from 500 psig to 3600 psig (CNG tank).

Since excess adsorption measures the difference between the mass of methane adsorbed and the mass of an equal volume of nonadsorbed methane, it follows that excess adsorption depends only on the surface area and how strongly the surface adsorbs methane, but not on the pore volume of the sample. For the total amount stored, Eq. (1), the situation is different: the volumetric storage capacity, $(m_{st}/m_s)\rho_a$, increases if the apparent density, ρ_a, increases,

$$\frac{m_{st}}{m_s}\rho_a = \left(\frac{m_{ads}^e}{m_s} - \frac{\rho_{gas}}{\rho_s}\right)\rho_a + \rho_{gas} \tag{2}$$

all other parameters being equal. Inversely, the gravimetric storage capacity, m_{st}/m_s, increases if the apparent density decreases, Eq. (1). Samples S-33/k and Batch 5.32 in Figure 3 conform perfectly with this structure-function relation: the DFT surface area is 2150 m²/g for both samples; the skeletal density is 2.0 g/cm³ for both samples; and excess adsorption is 193 g CH₄/kg carbon and 197 g CH₄/kg carbon, respectively. So the volumetric capacity of S-33/k is higher than that of Batch 5.32, and the gravimetric capacity is lower than that of Batch 5.32, because the apparent density of S-33/k is higher (Figure 5).

This observation indicates that the binding energy of CH₄ on the two samples is the same. The surface packing density, defined as excess adsorption divided by surface area [12], gives 0.0338-0.0345 CH₄ molecules per Å². Since excess adsorption is a lower bound to absolute adsorption (mass of adsorbed film), the reciprocal of the packing density gives an upper bound of 29-30 Å² for the surface area per CH₄ adsorption site on the two samples, at 35 bar and 293 K.

Hydrogen Storage

Sample S-33/k, which is our best-performing volumetric methane storage material (Figure 3), is simultaneously also our best hydrogen storage material (Figure 4 and Table II).

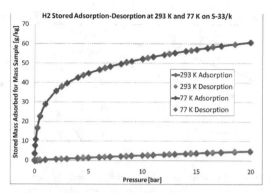

Figure 4. Gravimetric H_2 storage isotherm (total amount stored) on Sample S-33/k, a nanoporous carbon made from corncob. The excess adsorption isotherm was measured at Hiden Isochema, and converted into amount stored using Eq. (1) and the same structural data as in Figure 3.

Table II. Validation of H_2 storage results on S-33/k in three independent laboratories, and comparison with adsorbents in the literature (AX-21 is a commercial activated carbon; MOF-177 is a metal-organic framework). The values in the table are amount stored, and 79 g H_2/kg adsorbent is reported as 7.9 mass%. The values reported for Hiden are extrapolated from Figure 4.

	77 K, 47 bar	293 K, 47 bar	293 K, 80 bar
S-33/k, Hiden	7.9 mass%	1.2 mass%	1.9 mass%
S-33/k, U. Missouri	7.3-9.1 mass%	1.0-1.2 mass%	–
S-33/k, NREL	~8 mass%	1.4-1.6 mass%	2.1-2.4 mass%
AX-21 [13]	5.1 mass%	0.6 mass%	–
MOF-177 [14]	~10 mass%	~2.4 mass%	–

Surface and Pore-Space Characterization

Figures 5 and 6 collect structural data for the pore space of Sample S-33/k. The pore-size distribution shows that an extensive nanoporosity, bimodally peaked around 0.6 nm and 1.1 nm. The two peaks coincide with the optimum pore width of 0.6-0.7 nm and 1.1 nm predicted for maximum H_2 and CH_4 storage, respectively [4]. This may explain why S-33/k performs so well both for hydrogen and methane storage. The absence of any hysteresis in the N_2 adsorption and desorption isotherm, as well as transmission electron micrographs (not shown), confirm that S-33/k consist almost exclusively of pores less than 2 nm in width.

Figure 6 shows the SAXS data for S-33/k. SAXS is one of the few experimental methods that can "see" the spatial organization of pores across the entire solid, over 4 decades of length, 0.5-5000 nm. Structure at small length scales scatters at large wave vectors q, and structure at large length scales scatters at small q. Interestingly, the data at large wave vectors *cannot* be fitted by treating the pores in Figure 5a as independent scatterers. Instead, the "knee" in Figure 6 can be well fitted by the scattered intensity for cylindrical pores. The cylinder shape makes the pores correlated, and the best fit gives a pore width of 0.5 nm and a length of 1.4 nm [16]. This

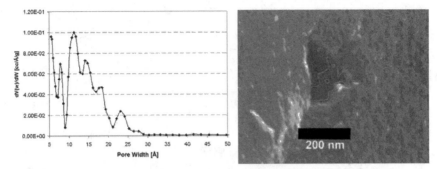

Figure 5. (a) Left: Pore-size distribution of S-33/k from N_2 adsorption. Most pores are in the range 0.5-1.0 nm. The volume of pores with diameter < 2.0 nm (micropore volume) is 1.2 cm^3/g. The DFT and BET surface area of the sample are 2150 m^2/g and 2500 m^2/g. Since the BET theory is not applicable to microporous systems [15], we quote BET areas just as figure of merit. (b) Right: Scanning electron micrograph of S-33/k, showing a rare large pore.

is in excellent agreement with the pore size data from N_2 adsorption (Figure 5a). At small q, the extended power law, $I \propto q^{-3.7}$, indicates the presence, at large length scales, of an external surface with fractal dimension ~2.3 [17], consistent with the mild roughness of the external surface visible in Figure 5b.

The extensive nanoporosity of S-33/k suggests that the surface has few graphitic domains. This is borne out by the Raman spectra of three different samples (Figure 7). The Raman spectrum of S-33/k is more consistent with an amorphous carbon structure than with a highly graphitic structure such as a carbon multi-wall nanotube (MWNT). The Raman spectrum of S-37 is more consistent with the spectrum that would be observed with a graphitic MWNT network [18]. It is possible that a surface richer in defects, which would result in a Raman spectrum

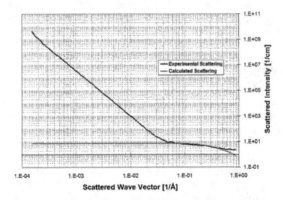

Figure 6. Small-angle x- ray scattering data of S-33/k. The red curve is the best fit of the knee at large scattered wave vectors, corresponding to cylindrical pores of width 0.5 nm and length 1.4 nm, consistent with the pore-size distribution in Figure 5a.

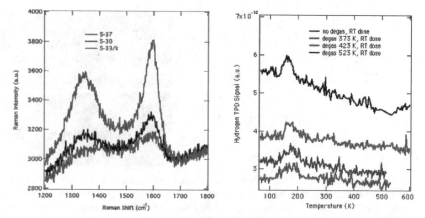

Figure 7. Raman spectra of three different samples (left), including S-33/k, and temperature-programmed H_2 desorption spectra of S-33/k under sequential degassing to 523 K (right).

similar to amorphous carbon, as in S-33/k, may have a higher H_2 adsorption capacity. However, the H_2 desorption peaks in the TPD spectra are all centered at ~150 K (Figure 7), consistent with that the surface of S-33/k physisorbs H_2 with a binding energy of ~4 kJ/mol, comparable to the binding energy on graphite. In pores of width 1.1.nm, this is not unreasonable, but in pores of width 0.6 nm, one would expect binding energies larger than 4 kJ/mol.

Computational work

Theoretical predictions of H_2 storage capacities under two distinct scenarios—localized and mobile adsorption—were carried out using the Langmuir adsorption isotherm and molecular dynamics simulations. Using the highly parallelized NAMD2 code [19], the simulations were carried out in a computational cell of size of approximately 100 Å × 100 Å × 100 Å, bisected by 6 layers of graphite, in which H_2 molecules interact via Lennard-Jones potentials with each other and with the carbon atoms in the graphite. A total of N = 1545, 3290, 4936, and 6581 H_2 molecules were used at each temperature, the time step for the integrations was Δt = 1 fs, and 100,000-200,000 time steps were used for each simulation. The sum of the Lennard-Jones potentials between a H_2 molecule and all carbon atoms gives an adsorption potential $V(x,y,z)$ with a strongly z-dependent attractive part and a weakly (x,y)-dependent corrugation part. Using the experimental binding energy of 5.0 kJ/mol for H_2 on graphite [20], we obtained approximately 0.5 kJ/mol for the peak-to-peak amplitude of the corrugation potential.

Simulated adsorption isotherms (Figure 8) were computed for various pairs (N, T), and equivalently for various pairs (P, T). From extrapolation of the number of molecules in the first layer at very high pressure, the surface area per adsorption site, $\alpha(T)$, was calculated. With appropriate values for the vibrational frequencies v_x, v_y, v_z of H_2 in the adsorption potential, Eqs. (4) and (5), this gave the framework to calculate the Langmuir isotherm,

$$\theta(p,T) = \frac{\chi(T)p}{1+\chi(T)p},$$ (3)

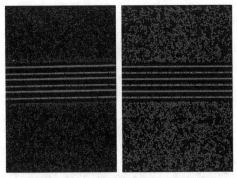

Figure 8. Snapshots of computer simulations showing adsorbed layers of H_2 on graphite at 77 K (blue) and 293 K (red). In both cases, the density of H_2 is noticeably higher near the graphite surface. The 6 layers of graphite creating the adsorption potential $V(x,y,z)$ are shown in green.

for surface coverage θ (number of adsorbed H_2 molecules per adsorption site, $0 \leq \theta \leq 1$), as a function of gas pressure p and temperature T. The expressions for the Langmuir constant $\chi(T)$ for localized and mobile adsorption are

$$\chi(T) = \frac{e^{E_B/(N_A kT)}}{\sinh(h\nu_x/(2kT))\sinh(h\nu_y/(2kT))\sinh(h\nu_z/(2kT))} \sqrt{\frac{h^6}{(8\pi m)^3 (kT)^5}}, \tag{4}$$

$$\chi(T) = \frac{\alpha(T)e^{E_B/(N_A kT)}}{\sinh(h\nu_z/(2kT))} \sqrt{\frac{h^2}{8\pi m(kT)^3}}, \tag{5}$$

respectively. In these expressions, $E_B > 0$ is the binding energy, defined as the depth of the minima of the potential energy $V(x,y,z)$, per mole of H_2; m is the mass of the H_2 molecule; N_A is Avogadro's constant; k is Boltzmann's constant; and h is Planck's constant. The surface area per site, $\alpha(T)$, is temperature-dependent because adsorbed molecules are densely packed at low temperature, but occupy a larger area at higher temperature due to thermal motion in the x, y direction. We note the elementary, but informative inequality $\alpha(T) <$ (surface packing density)$^{-1}$ from the discussion following Eq. (2).

Figure 9 shows excess adsorption isotherms calculated at 77 K and 293 K. The localized adsorption model at 77 K and the mobile adsorption model at 293 K, calculated with $E_B = 5.0$ kJ/mol, agree qualitatively well with experimental data over a broad of pressures. However, the Langmuir fit for 77 K falls below the experimental data at low pressures, consistent with the picture that sites with binding energies larger 5.0 kJ/mol, undoubtedly present in S-33/k, are occupied first and entail a higher coverage than with 5.0 kJ/mol. This agreement is remarkable in view of the experimental and theoretical input from widely different sources. Moreover, it allows us to discriminate experimentally between two vastly opposed situations of how molecules do or do not move along the surface, how such lateral dynamics (competition of vibrational and translational degrees of freedom) affects the H_2 storage capacity, and how control of such dynamics, e.g., by surface functionalization, may offer new venues to increase storage capacities by as much as a factor of two (ratio of mobile to localized adsorption at 77 K and ~40 bar in Figure 9).

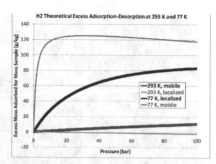

Figure 9. Theoretical H_2 excess adsorption isotherms computed from the Langmuir isotherm for localized and mobile adsorption. Localized and mobile adsorption at 77 K and 293 K, respectively (thick lines), are in good agreement with experimental data. Localized and mobile adsorption at 293 K and 77 K, respectively (thin lines), are ruled out by the experimental data. Excess adsorption drops with increasing pressure if the gas density increases more rapidly than the density of the adsorbed film, and turns negative (localized adsorption at 293 K) if the gas density exceeds the film density.

CONCLUSIONS

A novel class of carbons, made from waste corncob—a low-cost, renewable raw material, with superior storage capacities for natural gas (methane) and hydrogen has been presented. Volumetric and gravimetric methane capacities reported here, at room temperature, are 130% and 145%, respectively, of the best carbon in the literature we are aware of. The hydrogen capacities at cryogenic and room temperature are ~160% and 200%, respectively, of the best carbon we know of. Our cryogenic hydrogen capacity is about 80% of that of the best metal-organic framework. The natural-gas work is an integrated, lab-to-prototype RD&D effort. Remarkably, the best-performing material for natural gas is also a top performer for hydrogen storage. Evidence has been presented that this may be due to a bimodal pore-size distribution, with most pores having a width of less than 1.5 nm. Computational work has been presented that allows us to distinguish experimentally between localized and mobile hydrogen adsorption, that provides first evidence that the materials with high hydrogen storage capacity do carry a significant fraction of sites with high binding energies, and that offers design principles for surface-functionalized materials with improved hydrogen storage capacities.

ACKNOWLEDGMENTS

The authors thank McKinley Addy, Christian Bach, M. Frederick Hawthorne, Douglas B. Horne, Satish Jalisatgi, Mark W. Lee, Yun Liu, David F. Quinn, Francisco Rodríguez-Reinoso, Louis Schlapbach, Samuel C. Swearngin, and Andreas Züttel for valuable discussions. This material is based upon work supported by the National Science Foundation under Grant Nos. PFI-0438469 and DUE-0618459, U.S. Department of Energy under Award No. DE-FG02-07ER46411, U.S. Department of Defense under Award No. N00164-07-P-1306, U.S. Department of Education under Award No. P200A040038, University of Missouri (RB-06-40), and Midwest Research Institute. Acknowledgment is also made to the Donors of the American

Chemical Society Petroleum Research Fund (PRF43277–B5, M.W.R.). Use of the Advanced Photon Source was supported by the U.S. Department of Energy, Office of Science, Office of Basic Energy Sciences, under Contract No. DE-AC02-06CH11357. Special thanks are due to the Kansas City Office of Environmental Quality for lending us a NG vehicle of their fleet as test bed for the ANG tank.

REFERENCES

1. (a) M. Addy, T. Olson, and D. Schwyzer, "State Alternative Fuels Plan," California Air Resources Board and California Energy Commission, Publication No. CEC-600-2007-011-CTF, October 2007 (Sacramento, CA). http://www.energy.ca.gov/2007publications/CEC-600-2007-011/CEC-600-2007-011-CTF.PDF (b) M. Addy, P. Ward, and J. Wiens, "AB 1007 State Alternative Fuels Plan: Natural Gas Scenario," California Energy Commission and California Air Resources Board, May 31, 2007 (Sacramento, CA). http://www.energy.ca.gov/ab1007/documents/2007-05-31_joint_workshop/2007-05-31_NATURAL_GAS_SCENARIO.PDF (c) M. Addy, "Prospects, Challenges and Solutions for NGV's in a Low Carbon World," NGVAmerica Conference, October 16, 2007 (Reno, NV). http://www.cleanvehicle.org/conference/2007/presentations/3-Addy%20CEC%20Alt%20Fuel%20Plan%20Findings.pdf

2. U.S. Department of Energy and U.S. Department of Transportation, "Hydrogen Posture Plan—An Integrated Research, Development and Demonstration Plan," December 2006 (Washington, DC). http://www.hydrogen.energy.gov/pdfs/hydrogen_posture_plan_dec06.pdf

3. TIAX LLC, "Full Fuel Cycle Assessment: Well-to-Wheels Energy Inputs, Emissions, and Water Impacts—State Plan to Increase the Use of Non-Petroleum Transportation Fuels," Report to the California Energy Commission, Publication No. CEC-600-2007-004-REV, August 1, 2007 (Sacramento, CA). http://www.energy.ca.gov/2007publications/CEC-600-2007-004/CEC-600-2007-004-REV.PDF

4. (a) S.K. Bhatia and A.L. Myers, *Langmuir* 22, 1688 (2006). (b) A. Gigras, S.K. Bhatia, A.V.A. Kumar, and A.L. Myers, *Carbon* 45, 1043 (2007).

5. http://all-craft.missouri.edu

6. P. Pfeifer, F. Ehrburger-Dolle, T.P. Rieker, M.T. González, W.P. Hoffman, M. Molina-Sabio, F. Rodríguez-Reinoso, P.W. Schmidt, and D.J. Voss, *Phys. Rev. Lett.* 88, 115502 (2002).

7. T. Burchell and M. Rogers, *SAE Tech. Pap. Ser.*, 2000-01-2205 (2000).

8. P. Pfeifer, G.J. Suppes, P.S. Shah, and J.W. Burress, U.S. Patent Application No. 11/937,150 (November 8, 2007).

9. Y. Ginzburg, "ANG Storage as a Technological Solution for the 'Chicken-and-Egg' Problem of NGV Refueling Infrastructure Development." *Proceedings of the 23rd World Gas Conference* (International Gas Union, Amsterdam, 2006). http://www.igu.org/html/wgc2006/pdf/paper/add10822.pdf

10. (a) M. Eddaoudi, J. Kim., N. Rosi, D. Vodak, J. Wachter, M. O'Keeffe, and O. Yaghi, *Science* 295, 469 (2002). (b) T. Düren, L. Sarkisov, O.M. Yaghi, and R.Q. Snurr, *Langmuir* 20, 2683 (2004).

11. J.A. Chamot, *National Science Foundation, Press Release*, February 16, 2007, http://www.nsf.gov/news/news_summ.jsp?cntn_id=108390&org=NSF&from=news

12. (a) Y. Liu, H. Kabbour, and C.M. Brown, Abstract R2.3, 2007 MRS Fall Meeting, Boston. (b) Y. Liu, H. Kabbour, C.M. Brown, D.A. Neumann, and C.C. Ahn, *Langmuir*, in press (2008).

13. E. Poirier, R. Chahine, P. Bénard, D. Cossement, L. Lafi, E. Mélançon, T.K. Bose, and S. Désilets, *Appl. Phys. A* **78**, 961 (2004).

14. H. Furukawa, M.A. Miller, and O.M. Yaghi, *J. Mater. Chem.* **17**, 3197 (2007).

15. (a) P. Pfeifer and K.-Y. Liu, *Stud. Surf. Sci. Catal.* **104**, 625 (1997). (b) K.A. Sosin and D.F. Quinn, *J. Porous Mater.* **1**, 111 (1995).

16. M.B. Wood, J.B. Burress, J. Ilavsky, C.M. Lapilli, C. Wexler, and P. Pfeifer, to be published.

17. (a) H.D. Bale and P.W. Schmidt, *Phys. Rev. Lett.* **53**, 596 (1984). (b) P. Pfeifer and P.W. Schmidt, *Phys. Rev. Lett.* **60**, 1345 (1988).

18. A.C. Dillon, A.H. Mahan, P.A. Parilla, J.L. Alleman, M.J. Heben, K.M. Jones, and K.E.H. Gilbert, *NanoLetters* **3**, 1425 (2003).

19. J.C. Phillips, R. Braun, W. Wang, J. Gumbart, E. Tajkhorshid, E. Villa, C. Chipot, R.D. Skeel, L. Kale, and K. Schulten, *J. Comp. Chem.* **26**, 1781 (2005). See also http://www.ks.uiuc.edu/Research/namd/

20. L. Mattera, F. Rosatelli, C. Salvo, F. Tommasini, U. Valbusa, and G. Vidali, *Surf. Sci.* **93**, 515 (1980).

Mater. Res. Soc. Symp. Proc. Vol. 1041 © 2008 Materials Research Society   1041-R02-03

Hydrogen Adsorption in MOF-74 Studied by Inelastic Neutron Scattering

Yun Liu[1,2], Craig M. Brown[1], Dan A. Neumann[1], Houria Kabbour[3], and Channing C. Ahn[3]

[1]NIST Center for Neutron Research, 100 Bureau Drive, MS6102, Gaithersburg, MD, 20899

[2]Department of Materials Science and Engineering, University of Maryland, College Park, MD, 20742

[3]Division of Engineering and Applied Science, California Institute of Technology, Pasadena, CA, 91125

ABSTRACT

Adsorption of hydrogen and the occupancy of different binding sites as a function of hydrogen loading in MOF-74 are studied using inelastic neutron scattering (INS). Hydrogen molecules are observed to fully occupy the strongest binding site before populating other adsorption sites. The comparison of the INS spectra at 4 K and 60 K indicates that hydrogen adsorbed at the strongest binding site is strongly bound and localized. We also show that when two hydrogen molecules are adsorbed into a single, attractive potential well, the shortest inter-H_2 distance is about 3 Å, consistent with our previous observation of inter-H_2 distance when adsorbed in two neighboring potential wells.

INTRODUCTION

Adsorbing H_2 in nano-porous materials with large surface areas has attracted significant attention in past years. It has been shown for activated carbons that about 1 wt% of hydrogen can be adsorbed for every 500 m^2/g of surface area. [1] This has been a strong impetus in striving to increase material surface areas and achieve a concomitant increase in hydrogen uptake. For example, it has been reported that MOF-177 has N_2 Brauner-Emmett-Teller (BET) surface area of 4746 m^2/g; [2] and some carbon aerogels can have N_2 BET surface areas of ≈ 3200 m^2/g. [3] The saturation excess adsorption (SEA) at 77 K is ≈ 7.5 wt% for MOF-177 [2] and ≈ 5.3 wt% for carbon areogel. [3] However, due to the generally low binding strength of hydrogen, the SEA value decreases dramatically with increasing temperature. It has been shown in the cases of a few metal-organic framework (MOF) materials that the coordinately unsaturated metal centers (CUMCs) can greatly enhance the binding strength of H_2 in MOFs. [4-7] Of these examples, the Mn-BTT material has an initial hydrogen adsorption enthalpy of about 10.1 kJ/mol due to the direct interaction between hydrogen molecules and the exposed Mn^{2+} ions. [4] Hence, it is therefore extremely important to understand this hydrogen-CUMCs interaction and direct efforts for optimizing the material synthesis.

MOF-74 has one-dimensional pore channels of size ≈ 10 Å. [8-9] Previously, using neutron powder diffraction techniques, we have shown that the relatively large initial hydrogen adsorption enthalpy (about 8.8 kJ/mol) is due to a direct interaction of hydrogen molecules with the open Zn^{2+} ions in the framework. [7] The hydrogen molecules adsorbed on the surface of the pore are packed denser than the molecules in solid hydrogen. We have also shown, using a model system of two hydrogen adsorption sites placed close to each other, that the smallest interaction distance between physisorbed hydrogen molecules under technologically relevant conditions is only ≈3 A. [7] However, some issues remain unclear and will be addressed in this paper.

First, we performed site specific INS spectroscopy to understand how the different adsorption sites are occupied as a function of hydrogen loading. Comparison is made to the case of HKUST-1, where the strongest hydrogen adsorption sites are only partially occupied before weaker sites are populated. [10] Second, although we have demonstrated previously that the binding strength of the different sites in MOF-74 is significantly different, we would like to know how mobile the hydrogen molecules are with increasing temperature. This is achieved by comparing the INS spectra obtained at both high temperature and 4 K with exactly the same loading. Thirdly, we shall extend and compliment our previous calculations by looking at a system having a single potential well that is able to attract two hydrogen molecules.

EXPERIMENT

The synthesis and sample activation procedures for MOF-74 ($Zn_2(C_8H_2O_6)$) are already reported in previous publications. [7-9] Prior to the neutron experiments, the sample was further degassed at 120 °C overnight. The sample was then transferred into a vanadium sample can sealed with an indium o-ring and fitted with a valve. All the sample handling/transferring was performed in a helium glove box with oxygen and humidity sensors. The sample can was then mounted on a sample stick with a gas loading line. The gas loading line was first pumped to a good vacuum before opening the valve on the sample can. The sample stick was then put into a top-loading closed-cycle refrigerator (CCR) and cooled to 4 K for measurements.

Hydrogen gas at room temperature is a mixture of 25% para-H_2 and 75% ortho-H_2. [11] Although the effect of the mixture on hydrogen adsorption properties is not expected to be significantly in most situations, para-H_2 and ortho-H_2 have very large differences in neutron scattering cross sections. [12] In order to make the spectrum relatively simple, we have used a modified CCR with a paramagnetic catalyst to convert the normal hydrogen into predominantly para-hydrogen. Only para-H_2 was used in the experiments reported here.

For each hydrogen loading, the sample was first warmed to ≈ 50 K and then the calculated amount of H_2 gas was loaded to a container with a known volume at room temperature. The sample was then exposed to the hydrogen gas in the container. After the gas reached equilibrium at 50 K, The temperature was then slowly cooled down to 4 K. The pressure gauge always read zero before the temperature reached 25 K indicating all the hydrogen was adsorbed.

The INS spectra were obtained using the Filter Analyzer Neutron Spectrometer (FANS) at the National Institute of Standards and Technology Center for Neutron Research (NCNR). [13] The pyrolytic graphite monochromator was used with collimation of 20' and 20' before and after the monochromator, respectively, to produce a collimated, monoenergetic beam of neutrons. The overall energy resolution is about 1.2 meV between 6 meV and 15 meV and degrades with the further increase of energy.

DISCUSSION

Figure 1 shows the typical INS spectra of H_2 adsorbed in MOF-74 with loadings corresponding to 0.2 H_2:Zn, 0.6 H_2:Zn, 1.0 H_2:Zn, 1.2 H_2:Zn, and 2.0 H_2:Zn after subtracting the bare MOF-74 background spectrum. According to our previous work using neutron powder diffraction, there are 4 adsorption sites for hydrogen in MOF-74, with the first 3 adsorption sites dominating the adsorption properties at 77 K. [7] The first site is termed the 'Zn site' since the hydrogen molecule is directly associated with the Zn^{2+} ion. When there is only 0.2 H_2:Zn in the sample, we would expect that only the Zn site is occupied. Clearly, we can see that there are many peaks in the INS spectrum even when hydrogen molecules are adsorbed at one site. This seems to be a general observation for hydrogen molecules adsorbed in metal-organic frameworks [7,10] and hence, one needs to be careful in assigning peaks to different adsorption sites.

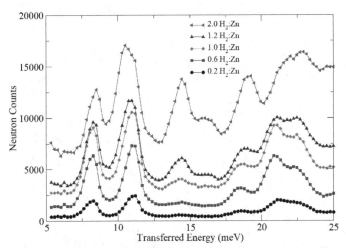

Figure 1. The inelastic neutron scattering spectra of adsorbed H_2 in MOF-74 after subtracting the bare MOF-74 spectrum. Error bars are smaller than the symbols.

H_2 consists of two protons with the spin ½ for each proton. The total wave function of the two fermions needs to be anti-symmetric. When the total spin of H_2, S, is 0, an H_2 is called para-H_2 and its rotational wavefunction have to be symmetric, i.e., the rotation quantum number, J, needs to be 0, 2, … . [11] When the total spin of the two protons, S, is 1, an H_2 is called ortho-H_2 and the associated J has to be 1, 3, … . For a freely rotating molecule in a three dimensional system, $E = BJ(J+1)$, where E is the rotational eigenenergy and $B = 7.35$ meV, the rotational constant of hydrogen. [11] Due to spin correlation, the neutron scattering cross section for transitions between para- states is very small. [12] On the other hand, the neutron scattering cross section of transitions between para- and ortho- states is very large and usually dominates an INS spectrum. [10, 12] Since we have used para-H_2 in these experiments, the prominent scattering features we observe are mainly due to transitions associated between para-H_2 and ortho-H_2.

When the spin state of a H_2 is changed, there is corresponding change of its rotational wavefunction, and hence a change of energy. For a freely rotating H_2, a peak at 14.7 meV is usually observed, which is due to the transition from $J = 0$ to $J = 1$. [11-12] At 4 K, all para-H_2 should be in the rotational ground state, i.e., $J = 0$. Because of the large rotational barrier of hydrogen adsorbed in MOF-74, the $J = 1$ eigenenergy levels are split and coupled with other rotational eigenstates and possibly translational motions, which result in the rich features for hydrogen adsorbed even at one adsorption site. A detailed analysis of the wavefunctions associated with each peak needs much more calculation, which will be addressed in a future paper. Here we take a phenomenological approach to understand the features of the spectra. Since the scattering cross section for a transition from para- to otho-H_2 is incoherent in nature, it reflects only the individual molecules. Hence the scattering features are additive. Therefore, by analyzing the intensity of each characteristic peak, we can estimate the amount of hydrogen adsorbed at the different sites.

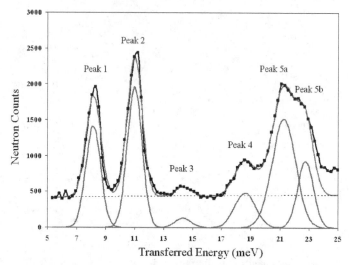

Figure 2. Gaussian peaks are used to fit the spectrum of 0.2 H_2:Zn loaded MOF-74. A slope function is used to fit the background signal. The FWHM of peak 1 and peak 2 are fixed as 1.2 meV, which is the instrument energy resolution in this energy range.

Figure 2 shows the spectrum for the case of 0.2 H_2:Zn along with a phenomenological description of the data. A slope function is used to simulate the background signal, which is shown as a dashed line. Each peak is fitted with a Gaussian function. The solid green line is the fitted curve. The full-width at half peak-maximum (FWHM) of peak1 at $E \approx 8.1$ meV and peak 2 at $E \approx 10.9$ meV is resolution limited and during the fitting the FWHM was fixed as 1.2 meV. Peak 1 and 2 are tentatively assigned to the rotational transitions associated with para- to ortho-H_2 transitions. The origin of peak 3, 4, 5 is difficult to be assigned without further measurements. Currently, we speculate that peak 3 may be due to the rotational translation coupling. Peak 5 contains at least two peaks and we used two Gaussian peaks (peak 5a and peak 5b) to fit it. One of them could be due to a rotational transition. The rotational peak splitting observed for H_2 adsorbed on single-walled carbon nanotubes or C_{60}, is only ≈ 1 meV. [14-17] The large rotational splitting observed here would be on the order of 10 meV, indicative of the much larger rotational barrier generated by the metal-organic framework. This picture is consistent with observation of H_2 adsorbed in HKUST-1. [10] When the amount of adsorbed hydrogen is larger than 1.0 H_2:Zn, peak 2 becomes broad and peak 3's intensity increases. Both peak 2 and 3 have to be fitted with multiple Gaussian peaks. This indicates that hydrogen molecules are adsorbed at different sites which have different rotational barriers, giving rise to the more complicated spectrum spectrum.

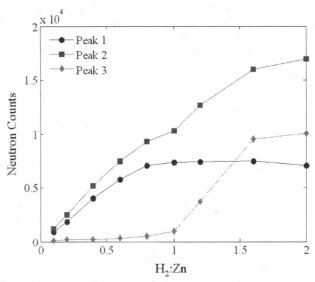

Figure 3. The variation of peak areas as a function of H_2 loading. The peak areas are obtained by fitting the peaks with one or two Gaussian functions. Peak 1 is a characteristic peak for H_2 adsorbed at site 1.

Figure 3 plots the neutron counts under each peak obtained by fitting the spectra as a function of adsorbed hydrogen. Peak 1 is a characteristic peak for the Zn site only. It saturates at ≈ 1.0 H_2:Zn. Peak 2 increases initially and begins to plateau at ≈ 1.0 H_2:Zn before the slope increases with further H_2 loading. This indicates that peak 2 is a characteristic feature of the Zn site at low loadings smaller than 1.0 H_2:Zn. When the Zn site is saturated, other sites begin to be populated causing the broadening and the increase in intensity of peak 2. This increase is mirrored with the change of intensity of peak 3. Initially, the intensity of peak 3 is very small. When the amount of adsorbed hydrogen is larger than 1.0 H_2:Zn, the intensity of peak 3 increases indicating that there are strong rotational transitions just under 15 meV due to population of sites other than the Zn site. We can therefore surmise that in MOF-74, the Zn site is first fully populated before other weaker binding sites are occupied. This is in contrast to what we observed in the case of HKUST-1, where the weaker binding sites are populated before the strongest binding site is fully occupied. [10] The possible reason for this behavior is that the binding energy difference between the first and subsequent sites in MOF-74 is much larger than that in HKUST-1.

Due to the strong binding of H_2 at the Zn site, H_2 molecules should not move too much even at high temperature. Figure 4 shows the INS spectra of H_2 adsorbed at the Zn site obtained at 4 K and 60 K. The sample was loaded with about 2.8 H_2:Zn and subsequently degassed for approximately 30 minutes at 60 K. After this time, we measured the INS spectra at both 4 K and 60 K. From comparison to the previously loaded data, we can estimate the amount of hydrogen as ≈ 0.6 H_2:Zn remaining after the degassing procedure. Comparing peak 1 and 2 at both temperatures, there is only a slight broadening indicating that the hydrogen molecules at the Zn sites are very strongly bound even at 60 K. The absence of any significant intensity at peak 3 indicates that the hydrogen molecules adsorbed at site 2 and site 3 are not so strongly bound that they can be removed at 60 K. Nevertheless, these sites can still be populated at 77 K when pressures are applied.

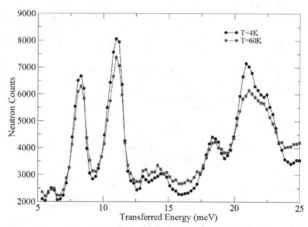

Figure 4. The comparison of the INS spectra of H_2 adsorbed in MOF-74 at 4 K and 60 K. Error bars are smaller than the symbols.

In the previous paper [7], we have estimated that for physisorbed hydrogen molecules pulled close to each other using strong attractive harmonic potentials, the smallest distance obtainable between two H_2 molecules is around 3.0 Å. However, sometimes, one harmonic potential may also hold multiple hydrogen molecules. Therefore, it is interesting to understand how close hydrogen molecules can approach each other under this situation. Two hydrogen molecules are put into a wide square well potential. A harmonic potential is then added, which is defined as

$$V_h(r) = \begin{cases} \dfrac{1}{2} m\omega^2 r^2 - E_d & \text{if } \dfrac{1}{2} m\omega^2 r^2 < E_d \\ 0 & \text{if } \dfrac{1}{2} m\omega^2 r^2 > E_d \end{cases}$$

where E_d is the potential depth and $\hbar\omega$ is the quantized energy difference between neighboring eigenenergy levels for a harmonic potential, m is the molecular mass of a hydrogen molecule. The inter-H_2 potential is taken from reference [11]. The average distance between the two hydrogen molecules, r_{HH}, is then calculated. The detailed calculation method is described elsewhere. [7]

The results are shown in Figure 5 upon changing the potential depth, E_d, and $\hbar\omega$. When $E_d = 0$ meV, r_{HH} is ≈ 4.9 Å. In general, with the increase of potential depth, r_{HH} decreases. At $\hbar\omega = 2$ meV, r_{HH} first decreases to a plateau at ≈ 4.58 Å at $E_d = 10$ meV, where only one H_2 molecule is held by the attraction potential and becomes less mobile. With a slight further increase of E_d, r_{HH} plummets to about 3.53 Å at E_d =15 meV because the potential well becomes wide enough for two H_2 molecules to be simultaneously accommodated. Once both H_2 molecules are attracted by the potential well, r_{HH} changes little with increasing E_d.

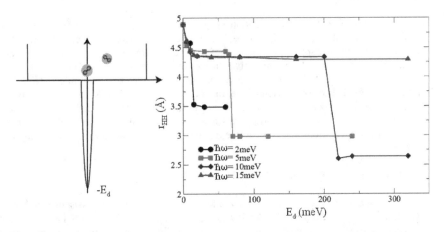

Figure 5. The calculation of the distance between two hydrogen molecules phsisorbed by one harmonic potential.

In order to move the H_2 molecules even closer, we have tested cases where $\hbar\omega$ = 5 meV, 10 meV, and 15 meV. With the increase of E_d, r_{HH} reaches a first plateau of about 4.4 Å, 4.3 Å, and 4.3 Å respectively. Larger $\hbar\omega$, indicating a steeper change of the potential well, results in a smaller r_{HH}. The relatively large r_{HH} shows that when only one H_2 is tightly bound to the attractive potential, the second H_2 can not approach very closely. With further increase of the potential depth, r_{HH} plunges to a second plateau with a value of r_{HH} about 2.94 Å for $\hbar\omega$ =5 meV and 2.6 Å for $\hbar\omega$ =10 meV when both H_2 are attracted. Once both H_2 molecules are in the potential well, an increase of $\hbar\omega$ decreases r_{HH}. Due to the strong repulsion between H_2 molecules, for a given $\hbar\omega$, E_d has to be large enough to hold both molecules inside the potential. There is no second plateau for $\hbar\omega$ =15 meV with potential depth up to 320 meV although we expect r_{HH} to reach a second plateau if the potential depth is further increased.

The results show that in order to bind H_2 molecules close to each other, a stronger binding potential is always necessary. Once both molecules are bound in a single potential, the steepness of the potential wall will eventually determine the equilibrium distance between H_2 molecules. We notice that the potential depth needed to bind two H_2 molecules is not too large. For $\hbar\omega$ = 5 meV, E_d only needs to be about 70 meV, which is readily reachable by most carbon based materials. In reality, the potential shape differs from a harmonic potential. The change of r_{HH} as a function of E_d may be smoother than that of the current case. When temperature is a factor, for instance at $77 K$, the thermal energy of a H_2 molecule is about 6.6 meV, which will affect the result of the second plateau when $\hbar\omega$ = 2 meV. For this case, a larger E_d will be needed to bind both H_2 molecules at $77 K$. However, the results for $\hbar\omega$ = 5 meV, 10 meV, and 15 meV are expected to be less affected. The minimum inter-H_2 distance can reach about 2.6 A when E_d is very large. For most carbon based materials, the surface could not provide such a strong potential, hence, the minimum distance that could realistically be achieved upon a carbon surface is only about 3 Å.

CONCLUSIONS

Hydrogen molecules adsorbed in MOF-74 have been studied using inelastic neutron scattering. By analyzing the different peak intensities due to hydrogen molecules adsorbed at different sites, the occupancy number at those sites are estimated as a function of adsorbed hydrogen. H_2 is observed to completely saturate the fist site before populating other sites, which is attributed to the large binding energy difference between the first adsorption site and other sites in MOF-74. The comparison of spectra obtained at 4 K and 60 K show that the first binding site strongly binds hydrogen even at 60 K. The estimation of the minimum hydrogen molecule distance based on surface adsorption is estimated to be around 3 Å, consistent with the results in our previous calculations and comparable to what we have previously measured using neutron diffraction.

ACKNOWLEDGMENTS

The authors wish to thank J. Leão and S. Slifer for experiment assistance and M. A. Green for useful discussion. This work was partially supported by the U. S. Department of Energy's Office of Energy Efficiency and Renewable Energy within the Hydrogen Sorption Center of Excellence.

REFERENCES

1. R. Chahine,P. Benard, In Advance in cryogenic engineering; Kittel, P., Ed.; Plenum Press: New York, 1998; Vol. 34, p 1257.
2. A. G. Wong-Foy, A. J. Matzger, O. M. Yaghi, *J. Am. Chem. Soc.* **128**, 3494-3495 (2006).
3. H. Kabbour, T. F. Baumann, J. H. Satcher, Jr., A. Saulnier, C. C. Ahn, *Chem. Mater.* **18**, 6085-6087 (2006).
4. M. Dincă, A. Dailly, Y. Liu, C. M. Brown, D. A. Neumann, J. R. Long, *J. Am. Chem. Soc.* **128**, 16876-16883 (2006).
5. M. Dincă, W. S. Han, Y. Liu, A. Dailly, C. M. Brown, J. R. Long, *Angew. Chem. Int. Ed.* **46**, 1419-1422 (2007).
6. V. K. Peterson, Y. Liu, C. M. Brown, C. J. Kepert, *J. Am. Chem. Soc.* **128**, 15578-15579 (2006).
7. Y. Liu, H. Kabbour, C. M. Brown, D. A. Neumann, C. C. Ahn, submitted (2007).
8. N. L. Rosi, J. Kim, M. Eddaoudi, B. Chen, M. O'Keeffe, O. M. Yaghi, *J. Am. Chem. Soc.* **127**, 1504-1518 (2005).
9. Rowsell, J. L. C., Yaghi, O. M. *J. Am. Chem. Soc.* **128**, 1304-1315 (2006).
10. Y. Liu, C. M. Brown, D. A. Neumann, V. K. Peterson, C. Kepert, *J. of Alloys and Compounds* **446-447**, 385 (2007).
11. I. F. Silvera, *Rev. Mod. Phys.* **52**, 393-452 (1980).
12. J. A. Young, J. U. Koppel, Phys. Rev. **135**, A603, (1964).

13. T. J. Udovic, D. A. Neumann, J. Leao, C. M. Brown, *Instrum. Methods* **A517**, 189 (2004).
14. C. M. Brown et al., *Chem. Phys. Lett.* **329**, 311 (2003).
15. Y. Liu et al. J. of Alloys and Compounds **446-447**, 368 (2007).
16. S. A. FitzGerald et al., Phys. Rev. B **60**, 6439 (1999).

Mater. Res. Soc. Symp. Proc. Vol. 1041 © 2008 Materials Research Society   1041-R02-04

Metal Hydrides for Hydrogen Storage

Jason Graetz, James J Reilly, and James Wegrzyn
Energy Sciences and Technology, Brookhaven National Laboratory, Upton, NY, 11973

ABSTRACT

The emergence of a Hydrogen Economy will require the development of new media capable of safely storing hydrogen with high gravimetric and volumetric densities. Metal hydrides and complex metal hydrides, where hydrogen is chemically bonded to the metal atoms in the bulk, offer some hope of overcoming the challenges associated with hydrogen storage. Many of the more promising hydrogen materials are tailored to meet the unique demands of a low temperature automotive fuel cell and are therefore either entirely new (e.g. in structural or chemical composition) or in some new form (e.g. morphology, crystallite size, catalysts). This proceeding presents an overview of some of the challenges associated with metal hydride hydrogen storage and a few new approaches being investigated to address these challenges.

INTRODUCTION

The transition to a Hydrogen Economy will require significant technological advancements in proton exchange membrane (PEM) fuel cells, hydrogen production and infrastructure. However hydrogen storage may be the most challenging technological barrier to the advancement of hydrogen fuel cell technologies for portable applications. At the heart of the issue is the low volumetric density of compressed H_2 gas. On a gravimetric scale one kilogram of H_2 can replace about 2.8 kg of gasoline (1 gallon). On a volumetric scale 1 gallon of gasoline is equivalent to about 12 gallons of high pressure H_2 gas (5000 psi).

The volumetric capacity is improved significantly (up to ~10×) by storing hydrogen in the solid state. This can be accomplished with adsorbents (e.g. activated carbon), where molecular hydrogen attaches to a surface through physisorption or absorbents (e.g. metal hydrides) where hydrogen disassociates and reacts with the solid in a chemisorption process. The interaction of hydrogen with light elements, such as aluminum and boron, has become an important research area in the energy sciences.

Table I. Overview of the three primary types of solid-state hydrogen storage in order of bonding strength showing the typical methods for hydrogen release and charging.

Media Type	Bonding Strength (ΔH_f)	Method of Hydrogen Release	Charging Method
Adsobents	Weak (<15 kJ/mol H_2)	Thermal decomposition (-200 - 50°C)	On-board (low temperature + high pressure)
Metal hydrides	Moderate (20-55 kJ/mol H_2)	Thermal decomposition (50 -200°C)	On-board (high pressure)
Ionic hydrides	Strong (>60 kJ/mol H_2)	Hydrolysis or high temperatures ($T > 200$°C)	Off-board (chemical regeneration)

This proceeding will briefly review a few of the reversible metal hydrides, some of the challenges associated with this approach and a few methods being explored to mitigate these problems. A less well-explored class of materials, the kinetically stabilized hydrides will be discussed as a possible alternative to the on-board reversible metal hydrides. Specific examples will be given for aluminum hydride (AlH$_3$). A few details on system design and regeneration will be presented for the kinetically stabilized hydrides. Although there is clear potential with these new materials, ultimately the true utility of these materials must be evaluated on a system basis rather than a materials basis. Therefore, much more system and life cycle analysis is necessary to understand strengths and weakness of different metal hydride hydrogen storage systems.

REVERSIBLE METAL HYDRIDES

In a typical reversible complex metal hydride the hydrogen atoms are associated with a metal in a complex (e.g. AlH$_6$, BH$_4$) and the hydrogen is released by a change in the thermodynamic conditions (decreasing P or increasing T). For a reversible material the hydrogenation reaction is exothermic, and therefore the release of H$_2$ requires the addition of energy (heat). For automotive applications with a PEM fuel cell the ideal source of this energy is the waste heat from the fuel cell. However, PEM fuel cells operate at low temperature and can supply heat at a temperature not much greater than 80°C. In the quest for new materials we are faced with two important constrains on the thermodynamics and kinetics of the hydrogenation/dehydrogenation reaction. For a 100 kW fuel cell the storage medium must supply H$_2$ at pressures greater than 1 bar (thermodynamic) and at a rate of around 2 gH$_2$/s (kinetic) at temperature of less than 80°C. At the present there are no known materials that have the desired high gravimetric capacity, thermodynamics and kinetics for this application and much of the research is focused on materials discovery and fundamentals of metal hydrides.

Recent Developments

Conventional metal hydrides that can readily supply hydrogen at room temperature typically have storage capacities < 2 wt. % [1] and cannot meet the demands of automotive PEM fuel cell applications. However, a number of complex hydrides have appreciable gravimetric hydrogen storage capacities as shown in Table 2. Although few of these materials are reversible, one of the best performing reversible metal hydrides is the catalyzed sodium alanate (NaAlH$_4$ + 2% Ti) [2]:

Table 2. Gravimetric hydrogen storage capacities of several metal hydrides.

Hydride	Total wt. % H
LiBH$_4$	18.5
Al(BH$_4$)$_3$	16.9
LiAlH$_4$	10.6
NaBH$_4$	10.5
AlH$_3$	10.0
Mg(AlH$_4$)$_2$	9.3
Ca(AlH$_4$)$_2$	8.2
NaAlH$_4$	7.5
KAlH$_4$	5.7
Mg$_2$FeH$_6$	5.5

$$3NaAlH_4 \leftrightarrow Na_3AlH_6 + 2Al + 3H_2 \leftrightarrow 3NaH + 3Al + 9/2H_2 \quad (1)$$

Despite the slow kinetics, the discovery of reversibility in the complex metal hydrides was a breakthrough in solid-state hydrogen storage. The catalyzed sodium alanate system exhibits approximately twice the reversible capacity of any of the conventional metal hydrides. Recent studies by Srinivasan et al. demonstrate reversible hydrogen cycling in Ti-catalyzed sodium alanate over 100 cycles with a measured capacity of nearly 4 wt. % H$_2$ at 160°C [3]. Although the capacity and cycling properties of catalyzed sodium alanate still do not meet the targets set by the U.S. Department of Energy (DOE), there is currently a flurry of activity looking at other high capacity complex metal hydrides that were historically viewed as irreversible.

Another recently developed reversible hydride that far exceeds the capacity of any of the conventional metal hydrides is the lithium amide/magnesium hydride system ($2LiNH_2 + MgH_2$) [4]:

$$Li_2MgN_2H_2 + 2H_2 \leftrightarrow Li_2MgN_2H_{3.2} + 1.4H_2 \leftrightarrow Mg(NH_2)_2 + 2LiH \qquad (2)$$

Recent studies have demonstrated greater than 4 wt. % reversible hydrogen cycling at 200°C with a capacity fade of $5x10^{-3}$ % per cycle or equivalently, 50% loss over 500 cycles (extrapolated) [5]. Similar to the sodium alanate system, the decomposition temperatures remain too high for low temperature fuel cell applications. Additional issues associated with the N-based hydrides include the tendency to release NH_3 during decomposition. The loss of N leads to capacity fade; and more importantly the presence of even trace amounts of NH_3 in the H_2 supply can poison the PEM fuel cell.

The previous examples of two of the best performing reversible metal hydride storage systems (catalyzed sodium alanate and lithium amide/magnesium hydride) demonstrate that we still have a long way to go. It is important to note that the discovery of a high capacity reversible hydride is simply the first step. The suitability of these new materials will be determined by their system performance and not their material performance. The few prototype tanks that have been developed have clearly shown that challenges with powder packing densities and heat exchangers may reduce the system capacity by at least 50% [6]. Although this area is emerging and we don't have a well-defined system or storage material there is a need for more life cycle analysis, safety, environment and engineering studies of new "candidate materials" to evaluate the performance of these new hydrides on a system level.

New Approaches

There are a number of known materials with high gravimetric hydrogen capacities (Table 2). However most, if not all of these materials exhibit unfavorable dehydrogenation thermodynamics. As previously mentioned, hydrogen must be released at low temperature (T $\approx$ 353 K) and useful pressures ($P_{353K} \approx$ 1-10 bar). The H_2 pressure at a given temperature is determined by the dehydrogenation thermodynamics. The decomposition enthalpy (ΔH) and entropy (ΔS) for a hypothetical material with ideal thermodynamics can easily be calculated using the van't Hoff equation:

$$P_{298K} = \exp\left[\frac{\Delta G_{f298K}}{RT}\right] = \exp\left[\frac{\Delta H_{f298K}}{RT} - \frac{\Delta S_{f298K}}{R}\right]. \qquad (3)$$

For hydrogenation of a metal using H_2 gas the entropy term (ΔS) is dominated by the entropy of the gas (S_{H2}=131 J/mol K). Therefore, the entropy is fixed (approximately) and the decomposition enthalpy of the ideal reversible material is around 39 - 46 kJ/molH_2. The thermodynamics of most hydrides fall outside this range and although most high capacity hydrides are too stable (evolve H_2 at high T) a few are too unstable (requiring extremely high pressures to rehydrogenate). Ideally, we could take a high capacity material and adjust the thermodynamics to fit our application.

One new approach to improving the dehydrogenation thermodynamics (i.e. reducing decomposition enthalpy) involves incorporating a second species into the reaction to stabilize the reaction product (Figure 1) [7]. The final state is stabilized by decomposing to an alloy (AH_x + $yB \rightarrow AB_y$ + $x/2H_2$) rather than the elements ($AH_x \rightarrow A$ + $x/2H_2$). Since the reaction now stops at a lower energy state the total enthalpy of the reaction is reduced and less energy (heat) is required. Vajo et al. have demonstrated this concept with a number off different hydride systems (LiH/Si, MgH$_2$/Si, LiBH4/MgH$_2$) [8]. Although this approach has proven to be effective, the addition of a second element (B) that does not contribute any new H$_2$ reduces the total gravimetric capacity of the system. In addition, this approach requires that the decomposition of the hydride (AH_x) and the alloy formation (AB_y) occur simultaneously in order to achieve the desired thermodynamic effect. Therefore, the key to this process is engineering the system with extremely fast kinetics. Much of the ongoing effort in this area involves improving the reaction kinetics by using nanocrystalline material incorporated into a scaffold [8] to increase surface area and avoid agglomeration.

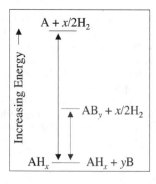

Figure 1. Energy diagram of the destabilization scheme [8] showing the original decomposition reaction (black) and the destabilized reaction (blue). The reaction enthalpy is reduced by stabilizing the final state as an alloy.

Another approach to tailoring the reaction thermodynamics involves engineering new less-stable metal hydrides. Many of the complex metal hydrides listed in Table 2 are much too stable and decompose at temperatures well above 100°C. The complex metal hydrides are made up of a metal hydride complex (e.g. AH$_6$, BH$_4$) and a metal cation (e.g. Li, Mg) that acts to stabilize the structure. One route to altering the decomposition thermodynamics involves substitutions of either the anion metal or the cation [9, 10]. For example, the hexahydride alanates tend to favor a structure with two different cation sites such as the fcc structure (Fm-$3m$) [9]. It was recently discovered that two different cations could be accommodated on these different sites to form the bialkali alanate M$_2$M'AlH$_6$ as shown in Figure 2a. Cation substitutions have a significant impact on the decomposition thermodynamics. An example is seen with the hexahydride alanate Na$_3$AlH$_6$, which has Na ions on both cation sites and an equilibrium H$_2$ pressure of 63 bar at 518 K. If a Na cation is replaced for a Li cation on the octahedral site the bialkali alanate Na$_2$LiAlH$_6$ is formed, which has an equilibrium pressure of 32 bar at 518 K. The equilibrium pressure is reduced by 50%. Although cation and anion substitutions can be used to control thermodynamics, presently only a few substitutions have proven to form stable compounds. A summary of the stable and unstable combinations for the bialkali alanates (cation substitution) is shown in Figure 2b.

The heat problem

One of the often-overlooked issues with the thermodynamics of metal hydride reactions is the vast quantity of heat generated during hydrogenation. Most research efforts that are focused on improving hydride thermodynamics are attempting to optimize the thermodynamics of the decomposition reaction. As previously mentioned, the ideal hydride has a decomposition enthalpy of around 39 - 46 kJ/mol H_2 (H_2 pressure of 1–10 bar at 353 K). Therefore, in a system that accommodates 5 kg of H_2 the total energy released during hydrogenation is around 100 MJ. Although this is a lot of wasted energy, the real problem arises when we consider the rapid refueling rates necessary to complete the process in approximately 3 minutes (1.67 kg H_2/min). Therefore, the 100 MJ of heat generated needs to be dissipated at a rate of 0.5 MW, which is an extremely challenging engineering problem that is not likely to be solved by moderate improvements in heat exchangers [11]. In all likelihood there are two ways around this problem, either significantly increase the time for recharging (probably not going to be a popular option) or simply swap the spent material with fresh material and allow the material to recharge slowly off-board. The importance of this point is that it opens the door to begin exploring new materials that have traditionally been overlooked since they are not easily reversible on-board.

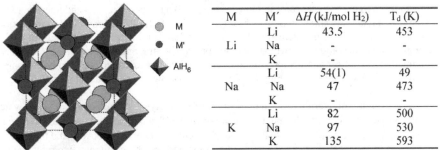

M	M′	ΔH (kJ/mol H_2)	T_d (K)
	Li	43.5	453
Li	Na	-	-
	K	-	-
	Li	54(1)	49
Na	Na	47	473
	K	-	-
	Li	82	500
K	Na	97	530
	K	135	593

Figure 2: (a) Structure of bialkali alanate $M_2M'AlH_6$ showing large (yellow) and small (green) alkali metal cations along with AlH_6 octahedra (blue). **(b)** Decomposition enthalpy and temperature for the bialkali alanates is based on experimental isothermal data [9].

KINETICALLY STABILIZED HYDRIDES

The kinetically stabilized hydrides represent a class of materials that have received little recent attention due to poor reversibility, but may offer some advantages over the well-known reversible hydrides. The kinetically stabilized hydrides are characterized by a low reaction enthalpy (< 30 kJ/mol) and are metastable under ambient conditions. Although the decomposition reaction is thermodynamically favorable at room temperature, H_2 evolution is limited by the kinetics. A few examples of kinetically stabilized hydrides include AlH_3, $LiAlH_4$, Li_3AlH_6, $Mg(AlH_4)_2$, $Ca(AlH_4)_2$, among others. Although the mechanism(s) responsible for stabilizing these materials is not well understood, a few likely possibilities include slow hydrogen/metal diffusion and surface barriers that inhibit the formation of molecular hydrogen. One advantage of using a metastable hydride is the low reaction enthalpy, which reduces the heat required to release the hydrogen at practical pressures. In addition, these materials exhibit rapid low temperature (80–100ºC) H_2 evolution rates, due to the large driving force for decomposition. Along with the thermodynamic and kinetic advantages, the kinetically stabilized hydrides also present a new set of challenges. The first issue is engineering a material and system that has good control over the decomposition reaction so that there are no runaway reactions and no unwanted H_2 evolution. Secondly, and probably more importantly, these materials cannot be formed by direct hydrogenation and therefore new energetically efficient methods to regenerate or recycle these hydrides directly from the spent material and hydrogen gas need to be developed.

Aluminum hydride

One of the more promising kinetically stabilized hydrides for low temperature PEM fuel cell applications is aluminum hydride (AlH_3). AlH_3 has a volumetric hydrogen capacity of 148 g/L (greater than twice that of liquid hydrogen!) and a gravimetric hydrogen capacity exceeding 10 wt. %. AlH_3 evolves hydrogen gas at low temperatures (< 150°C) via the following decomposition reaction:

$$AlH_3 \rightarrow Al + 3/2H_2. \tag{3}$$

In addition to the high capacity and low decomposition temperature, this endothermic reaction requires a relatively small amount of heat (7 kJ/mol H_2 [12]). Therefore, the decomposition reaction can easily be maintained by heat from the fuel cell. In fact, this material can be engineered to decompose slowly at room temperature (extracting ambient heat). A number of attractive low-power applications (remote sensors, data transmission, etc.) can be envisaged for AlH_3, particularly those involving portable or remote power systems using a small hydrogen powered fuel cell.

Aluminum hydride forms a number of different crystallographic phases, all of which are thermodynamically unstable under ambient conditions. Thermal stability studies of α, β and γ-AlH_3 [12] were used to calculate the equilibrium pressures between α-AlH_3 and the elements (Al + H_2) at various temperatures. These results, combined with high-pressure experimental results [13], give a H_2 equilibrium pressure at room temperature of approximately 7 kbar [14]. Although extremely high pressures are required to thermodynamically stabilize this material, large crystallites of AlH_3 are extremely stable and can be stored in air for thirty years with little to no decomposition [11].

Controlling decomposition and system design

The ability to control the decomposition rate and engineer a material that exhibits the full spectrum of H_2 rates (0 – 0.02 $gH_2s^{-1}kW^{-1}$) within the appropriate temperature window (40 – 150°C) is crucial for any system based on a kinetically stabilized hydride. Hydrogen storage systems based on the conventional metal hydrides control the rate of decomposition through the tank pressure. However, for a metastable hydride like AlH_3, the equilibrium pressure is much too high to limit decomposition and the release of H_2 must be controlled thermally. Therefore, new efforts are needed to engineer materials for a thermally controlled hydrogen storage system.

Aluminum hydride exhibits rapid hydrogen evolution rates at low temperature [15], but can this material meet the demanding DOE full flow rate target of 0.02 (gH_2/s)/kW [16]? Our recent kinetic studies evaluated the isothermal H_2 evolution rates in three phases α, β and γ-AlH_3

at temperatures 60–140°C [17]. We demonstrated that 100 kg of fine crystalline α-AlH₃ (0.1 μm) will meet the full flow target of 1.0 g H₂/s (50 kW fuel cell) at a temperature of 115°C (Figure 3a and 3b) [15, 17]. This study also demonstrated that the stability and decomposition rates vary significantly with crystallite size (Figure 3). The fine crystallites (0.1 μm) of α-ALH₃ (shown in Figure 3b) exhibit a decomposition temperature below 60°C with a rate of 1.0 g H₂/s at around 115°C. The material composed of 50 μm crystallites (large cuboids shown in Figure 3c) decomposes at temperatures above 130°C with a H₂ evolution rate more than 2 orders of magnitude slower. These results suggest that the temperature-dependent H₂ rates can be tuned by controlling the crystallite size. A better understanding of the role of size and morphology will be critical for the development of a kinetically stabilized hydrogen storage system.

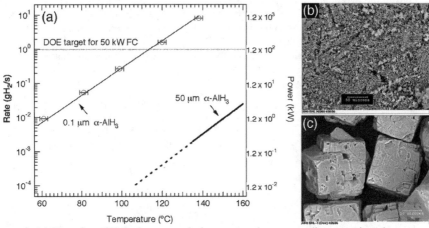

Figure 3. (a) Plot of α-AlH₃ hydrogen evolution rates and corresponding scanning electron microscopy images for two different crystallite sizes: (**b**) 0.10 μm [5] and (**c**) 50 μm [8]. Rates are based on 10 kg H₂ (100 kg AlH₃) and power values were determined from the lower heating value of H₂ (120 kJ/g H₂). The red line represents the maximum fuel flow target 1.0 g H₂/s for a 50 kW fuel cell.

We have demonstrated that α-AlH₃ can meet the full flow rate for a small fuel cell, but once the reaction has been initiated can the H₂ evolution be stopped? To answer this question decomposition rates were investigated by isothermal and intermittent isothermal decomposition. The plot in Fig. 4 shows the total H₂ evolved (red circles) and the first derivative of the total H₂ evolved (blue line), which corresponds to the H₂ rate for 100 kg of material. We found that the H₂ evolution rate can be slowed and even stopped by reducing the reactor temperature to room temperature. When the reactor is heated back to the decomposition temperature (120°C), the rate returns to the rate before the temperature was reduced. In fact, regardless of whether the decomposition occurs in one step (Figure 4a) or after multiple starts on stops (Figure 4b) the general Lorentzian shape of the rate curves are strikingly similar. This suggests that the full spectrum of H₂ evolution rates (e.g. 0.0-1.0 gH₂/s) can likely be obtained with a 100 kg α-AlH₃ with a variable temperature hydride bed operating at 23 ≤ T ≤ 115°C. However, heating and cooling a large thermal mass such as a hydride bed is not likely to be the design. A more sensible approach may be to use a pumpable slurry where the AlH₃ is transferred to a reactor, decomposed, and returned to separate region of the tank for spent fuel.

Regeneration

Direct hydrogenation of Al metal requires extremely high pressure and therefore AlH₃ is typically prepared by an organometallic reaction [18]. Although there are a number of different possible methods to regenerate AlH₃, one of the easiest may be to simply recycle the byproducts (Al and LiCl) as shown in Figure 5. This is a hypothetical "best case" route, which utilizes the formation energy of the AlH₃ precursors (AlCl₃ and LiH) in the splitting of the LiCl. In this method a minimum energy input of 167 kJ/mol H₂ (70% of the fuel energy) is necessary to convert the byproducts into the precursors for forming AlH₃. Although this process is costly it may be more energetically efficient than other low temperature hydrogen storage methods that utilize a hydrolysis reaction to release H₂. The byproducts of these reactions are left in a deep thermodynamic well as oxides and hydroxides. In comparison, AlH₃ can be thermally decomposed at low temperature directly to Al metal and 10.1 wt.% H₂ with much simpler regeneration methods at comparable cost.

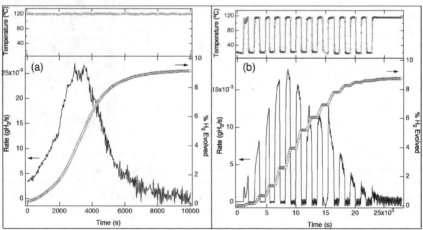

Figure 4. (a) Isothermal and (b) intermittent isothermal decomposition at 120°C showing the percent H₂ evolved (red circles), the rate of H₂ evolution based on 1 kg of AlH₃ (blue line) and the sample temperature (above, green).

Other approaches to the regeneration of AlH₃ under mild pressure and temperature conditions involve the formation of stable intermediate phases, such as alane adducts. Recently, Reilly et al. discovered a reversible hydrogenation reaction using Ti-catalyzed Al powder and triethylenediamine (TEDA) in THF to form the alane adduct (AlH₃-TEDA) at pressures less than 20 bar [19]. The AlH₃-TEDA product is insoluble in THF and precipitates from solution. Implicit in this regeneration scheme is a second step whereby the TEDA is removed and the pure, crystalline AlH₃ is recovered. Although this separation step may be difficult for the stable AlH₃-TEDA adduct, a number of other alane amines and alane adducts can be envisaged that may be more amenable to separation.

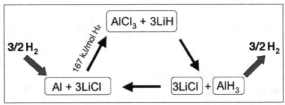

Figure 5. AlH₃ fuel cycle showing the synthesis from AlCl₃ and 3LiH (top), which produces the fuel AlH₃ and the byproduct LiCl (right). After the fuel is consumed (H₂ removed) the spent Al is recombined with the LiCl in the recycling step (left), which requires a minimum energy 167 kJ/mol H₂ to return to the original precursors.

CONCLUSIONS

There are currently no hydrogen storage materials that come close to meeting all of the DOE system targets. Although basic materials development and hydride research has led to significant improvements in the metal hydride properties (thermodynamics and kinetics), much more is needed to meet the DOE goals. Onboard hydrogenation will be extremely challenging in an acceptable time period (3-5 minutes). The kinetically stabilized metal hydrides with their low ΔH and rapid low temperature kinetics offer some new hope, but also present a number of new challenges. New concepts are needed for systems based on kinetically stabilized hydrides and new studies are needed to investigate hydride slurries, off-board refueling, low temperature decomposition (shelf life) and safety. In addition, new cost effective and energy efficient regeneration processes need to be developed to recycle the metastable hydrides from the spent material and H₂ gas.

ACKNOWLEDGMENTS

This work was supported through the Metal Hydride Center of Excellence (Office of Energy Efficiency and Renewable Energy) and the Hydrogen Fuel Initiative (Office of Basic Energy Sciences), U.S. Department of Energy under Contract No. DE-AC02-98CH1-886.

REFERENCES

1. G. Sandrock, *J. Alloys Compd.* **293-295**, 877 (1999).
2. B. Bogdanovic and M. Schwickardi. *J. Alloys Compd.* **253–254**, 1 (1997).
3. S. S. Srinivasan, H.W. Brinks, B.C. Hauback, D. Sun and C.M. Jensen, *J. Alloys Compd.* **377**, 283 (2004).
4. W. Luo and S. Sickafoose, *J. Alloys Compd.* **407**, 274 (2006).
5. J. Wang, DOE Metal Hydride Center of Excellence Kick-Off Meeting, Livermore, CA, 2005.
6. D. Mosher, DOE Hydrogen Program: 2006 Annual Progress Report – Storage, p, 281, http://www.hydrogen.energy.gov/annual_progress06_storage.html.
7. J. J. Vajo, F. Mertens, C. C. Ahn, R. C. Bowman Jr. and B. Fultz, *J. Phys. Chem. B 108*, 13977 (2004).

8. J. J. Vajo, G. L. Olson, *Scripta Materialia* **56**, 829 (2007).

9. J. Graetz, Y. Lee, J. J. Reilly, S. Park and T. Vogt, *Phys. Rev. B* **71**, 184115 (2005).

10. E. Ronnebro and E. H. Majzoub, *J. Phys. Chem. B* **110**, 25686 (2006).

11. G. Sandrock, J. Reilly, J. Graetz, W.-M. Zhou, J. Johnson and J. Wegrzyn, *Appl. Phys. A* **80**, 687 (2005).

12. J. Graetz and J. J. Reilly, *J. Alloys Comp.* **424**, 262 (2006).

13. A. K. Konovalov and B. M. Bulychev, Inorg. Chem. **34**, 172 (1995).

14. J. Graetz, S. Chaudhuri, Y. Lee, T. Vogt and J.J. Reilly, *Phys. Rev. B* **74**, 214114 (2006).

15. J. Graetz and J.J. Reilly, J.G. Kulleck and R. C. Bowman, Jr. *J. Alloys Comp.* **446-447**, 271 (2007).

16. DOE Hydrogen, Fuel Cells & Infrastructure Technologies Program Multi-Year Research, Development, and Demonstration Plan, Hydrogen Storage Technical Plan, 2007 http://www1.eere.energy.gov/hydrogenandfuelcells/mypp/

17. J. Graetz and J. J. Reilly, *J. Phys. Chem. B* **109**, 22181 (2005).

18. F.M. Brower, N.E. Matzek, P.F. Reigler, H.W. Rinn, C.B. Roberts, D.L. Schmidt, J.A. Snover, K. Terada, *J. Am. Chem. Soc.* **98**, 2450 (1976).

19. J. Graetz, S. Chaudhari, J. Wegrzyn, Y. Celebi, J.R. Johnson, W. Zhou, J.J. Reilly, *J. Phys. Chem. C* **111** 19148 (2007).

Mater. Res. Soc. Symp. Proc. Vol. 1041 © 2008 Materials Research Society   1041-R02-05

Science and Prospects of Using Nanoporous Materials for Energy Absorption

Xi Chen[1], and Yu Qiao[2]
[1]Columbia University, New York, NY, 10027
[2]University of California, San Diego, La Jolla, CA, 92093

ABSTRACT

With its ultra-large specific surface area, a nanoporous material is an ideal, yet relatively unexplored, platform for accepting or actuating liquids, with potential performance gains for energy dissipation and output typical of disruptive technologies. Our experimental and theoretical results indicate either dramatically improved performance or unique combinations of properties and capabilities not attainable in conventional materials, which make the novel nanoporous structures studied herein very attractive as advanced protective intelligent systems.

INTRODUCTION

For more than a decade, intensive studies have been conducted on manufacturing nanostructured protective composites. The basic idea is quite straightforward: in a nanoparticle or nanolayer reinforced composite, if under external loadings most nanofillers could debond from the matrix, a large amount of energy would be dissipated at the nanofiller-matrix interface - such a composite would be ideal for protection or damping applications, including car bumpers, body armors, mounting stages, etc. However, one intrinsic difficulty is the lack of control of the filler-matrix interaction. For example, very often the nanocomposite becomes less ductile with the addition of the nanofillers; thus, it tends to fail by catastrophic cracking and only a small fraction of nanofiller-matrix interface could be involved in the energy dissipation process, i.e. the large interface area cannot be utilized.

In order to fully take advantage of the large surface/interface area of a nanostructured system, at least one of the components must be sufficiently "flexible". Inspired by this understanding, and noticing the fact that the most "flexible" materials are actually liquids, recently we have pioneered incorporating liquid phases in nanoporous materials.

Nanoporous materials are solids containing large volume fractions of nano-sized pores (Fig. 1). The most dominant characteristic of them is the ultrahigh specific area of pore surfaces. They have been widely used for absorption and catalysis, yet the potential applicability in mechanical systems has not received the necessary attention. The system under our investigation is manufactured by dispersing surface charged nanoporous particles in a nonwetting liquid. Beyond a critical pressure, the "flexible" liquid phase could be forced into the nanopores and almost all of the nanopore surfaces could be exposed to it. Thus, the solid-liquid interactions (i.e. the capillary effect) can be greatly amplified by the large specific area, A. That is, accompanied by the pressure-induced infiltration, a large amount of external work would be transformed into the solid-liquid interfacial energy, which can be regarded as being absorbed. Denoting $\Delta\gamma$ as the excess solid-liquid interfacial

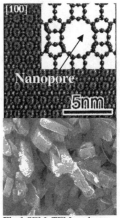

Fig.1 SEM, TEM, and pore structure of synthetic nanoporous zeolites.

tension, as a zero order approximation, the system energy density can be envisioned as $\Delta \gamma A$. Since A is 100-2000 m^2/g for nanoporous materials (compared to ~0.0001 m^2/g for bulk materials), the energy absorption efficiency of this system can be much higher than that of ordinary protective materials.[*] Moreover, since the pore size is comparable with the solid-liquid interface zone thickness, the reorganization of the liquid molecules in the interface zone is dominant to the "flow" of the confined liquid, resulting in additional beneficial internal friction for energy dissipation.

The seamless combination of the nm-level size effect and the full use of large specific surface area also enables eminent intelligent actuation, since the behaviors of the confined liquid in nanopores can also be controlled thermally or electrically. As temperature or voltage varies, due to the well-known thermocapillary or electrocapillary effect, the solid-liquid interfacial tension (i.e. the wettability) would vary considerably. Consequently, the liquid tends to "flow" into or out of the nanopores, leading to a significant reversible system volume variation as a voltage is applied or the temperature is adjusted. Once again, thanks to the large specific area, A, the system energy density, which at the order-of-magnitude level can be estimated as $\delta \gamma A$, would be higher than that of conventional smart solids such as Ti-Ni alloys and piezoelectrics by orders or magnitude, with $\delta \gamma$ being the averaged effective variation in $\Delta \gamma$ caused by the temperature/voltage change. The deformability is determined by the porosity, usually 30-60%. In this short paper, we focus on the energy absorption of NEAS.

EXAMPLE OF NEAS AND ITS PERFORMANCE

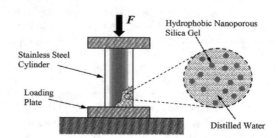

Fig. 2 Schematic of NEAS enhanced steel cylinder

Honeycomb is one of the most widely applied cellular energy absorption structures. While it can absorb energy via plastic buckling, it is not reusable; moreover, the characteristic buckling time depends on the structural dimension, which can be of macroscopic scale and thus the response time may be too slow for useful energy absorption. One method to solve this problem is to reinforce the honeycomb by the novel NEAS whose response time is on the order of microseconds. The experimental setup depicted in Fig.2 was developed. The testing system consisted of a thin-walled stainless steel cylinder and a liquid phase sealed in it. The height, outer diameter, and wall thickness of the cylinder were $h = 25.4$ mm, $2R = 6.86$ mm, and $t = 0.13$ mm, respectively. Both ends of the cell were fixed on stainless steel loading plates using a J-B Weld

---

[*] For instance, if $\Delta \gamma A = 100$ J/g ($\Delta \gamma$ is typically 30-300 mJ/m^2), the kinetic energy of a 2000-lbs car at 70 MPH can be entirely absorbed by only about 10 lbs of nanoporous material, with a liquid phase about 10-15 lbs in weight. Note that currently a commercial car bumper would fail to protect the vehicle once the speed exceeds 20 MPH.

epoxy glue. The cell was filled by an aqueous suspension of 0.4 g of Fluka 100 C_8 reversed phase nanoporous silica gel. The sample preparation was performed under water so that no air was entrapped. The stainless steel cylinder was employed as an analog to an individual cell at the lateral surface of a honeycomb panel.

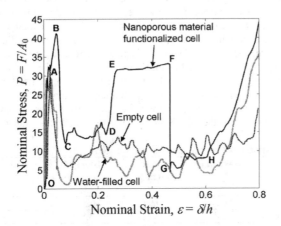

Fig.3 The typical stress-strain curves. The nominal stress is defined as $P = F/A_0$, and the nominal strain is $\varepsilon = \delta/h$, where F is the applied compressive load, A_0 is the cross-sectional area of the cell, δ is the displacement, and h is the initial cell height.

The nanoporous silica gel was hydrophobic, with the surface coverage of about 12%. The average nanopore size was 7.8 nm, with the standard deviation of 2.4 nm. The specific nanopore volume was 0.55 cm^3/g. At the infiltration pressure of about 18 MPa, one gram of the silica gel can dissipate about 14 J of energy. More details of the NEAS are documented elsewhere [1,2].

The prepared samples were tested using a type 5569 Instron machine. The compressive load, F, was applied and measured by an Instron 50KN loadcell. The displacement, δ, was measured by a linear variable displacement transducer. The crosshead speed was set to 0.5 mm/min. For comparison purpose, empty cells and cells filled by distilled water were also tested. For each type of cell, 3-5 samples were analyzed. Figure 3 shows typical stress-strain curves, where the nominal stress is defined as $P = F/A_0$, and the nominal strain is defined as $\varepsilon = \delta/h$, with $A_0 = 36.9$ mm^2 being the cross-sectional area of the steel cell [3].

It can be seen clearly that the deformability of the empty cell is quite high. After the initial linear compression stage, as the buckling initiates and continues, a broad plateau is formed, with the width of more than 70% of the initial cell height. As discussed above, due to the change in cell wall configuration, the buckling initiation stress is much higher than the buckling development stress, i.e. the height of the plateau is considerably lower than the height of the first peak. The buckling plateau is quite jerky, consisting of a number of "bumps". Each "bump" reflects the stress accumulation, formation, and folding process of a wrinkle. As most of the cell wall is folded, the structure is no longer compliant and the nominal stress rises rapidly as the displacement further increases.

As the cell is filled by distilled water, since the compressibility of water is negligible, the cell-wall buckling is suppressed, and initially the energy dissipation is dominated by the extended cell-wall yielding along the radius direction caused by the inner pressure. As a result, the system becomes quite rigid, and shortly after the peak load is reached, the inner pressure becomes sufficiently high such that abrupt cracking takes place along the longitudinal direction. After the liquid leaks, the system behavior resembles that of an empty cell, except that, since the longitudinal crack weakens the cell wall, the buckling of the fractured cell occurs at a relatively low stress level. Depending on the crack length, the decrease in buckling stress is in the range of 5-25%. Therefore, the energy absorption capacity is reduced, as shown in Table 1, where the absorbed energy, U, is calculated as the average area under the load-displacement curves in the nominal strain range of 0 to 0.75. Clearly, both of the energy absorption efficiencies per unit mass, U/m, and per unit volume, U/V, are smaller than that of the empty cells, with m and V being the system mass and the system volume, respectively. For the empty cell, $U/m = 11.5$ J/g, slightly lower than that of the NEAS. For a water filled cell, this value decreases by nearly 70%.

Table 1. Comparison of empty, water-filled, and NEAS-filled cells.

	Mass, m (g)	Volume, V (cm³)	Absorbed Energy, U (J)	U/m (J/g)	U/V (J/cm³)
Empty Cell	0.57	0.94	6.54	11.5	6.96
Water Filled Cell	1.48	0.94	5.80	3.92	6.17
NEAS Filled Cell	1.20	0.94	16.7	13.9	17.8

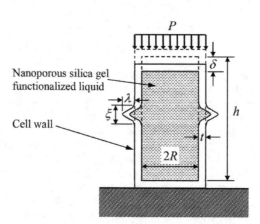

Fig.4 A schematic diagram of the buckling initiation process of a NEAS enhanced cell. The dashed lines indicate the initial configuration, and the solid lines indicate the deformed configuration.

The solid curve in Fig.3 indicates the behavior of a cell filled by the NEAS. Initially, in the linear compression stage ("OA"), the system behavior is quite similar with that of the empty cell and the water filled cell, as it should be. As the critical nominal stress of about 35 MPa is reached, at point "A", the cell wall deforms plastically along the radius direction, as the pressure in the liquid phase is insufficient to trigger the pressure induced infiltration and therefore the buckling is suppressed. As the cell is compressed, the pressure increases continuously. Eventually, at point "B", the pressure induced infiltration in the largest nanopores is activated and the liquid phase becomes highly compressible. As a result, buckling takes place. As a major wrinkle is formed, the stress quickly drops to "C". In an empty cell, the cell wall can buckle either inward or outward. In the NEAS filled cell, however, since the cell wall is supported by the liquid phase, the buckling can occur only outward, leading to a sudden increase in system volume, ΔV, as depicted in Fig.4. Assume that the wrinkle is sigmoidal, the value of ΔV can be estimated as $4\xi\lambda R$, where λ is the magnitude of the wrinkle. According to the experimental measurement, the wrinkle wavelength is around 2.5 mm, and the magnitude is around 0.5 mm. Thus, ΔV is 17.2 mm^3. Due to the expansion in volume, the cell becomes only partly filled. Under this condition, the system behavior is similar to that of an empty cell, until the cell is compressed to point "D" and ΔV is "consumed" by wrinkle folding. The volume decrease associated with "CD" is about 140 mm^3, which is much larger than the calculated ΔV, probably due to the continuous plastic expansion of the cell wall.

As the liquid phase starts to carry load again, the nominal stress increases from "D" to "E". The effective stiffness of the system in this section is about the same as that in the initial linear compression stage, which is dominated by the modulus of elasticity of cell wall and the bulk modulus of water. When P rises to 32 MPa, the pressure induced infiltration is re-activated, forming the second plateau "EF". As the plastic strain in cell wall is increasingly large, abrupt cracking occurs at point "F", somewhat similar with the failure observed in the water-filled cell. The total system volume change associated with the second loading is about 170 mm^3, smaller than the total nanopore volume; that is, the energy absorption capacity of the nanoporous silica gel is not fully utilized. As the cell becomes empty, the nominal stress drops to 7 MPa, after which the system behavior is similar to that of an empty cell (section "GH").

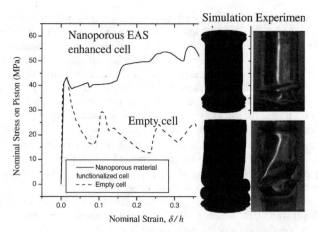

Fig.5 Simulation of the compression test of a steel tube containing NEAS vs. an empty tube. The large area below the NEAS enhanced cell demonstrates exceptional energy absorption capability.

The energy absorbed by the NEAS enhanced cell is 16.7 J. Hence, the mass based energy absorption efficiency, U/m, is 13.9 J/g, and the volume based energy absorption efficiency, U/V, is 17.8 J/cm^3. Compared with the energy absorption efficiency of an empty cell, the former is more than 20% larger and the latter is more than two times higher. Clearly, using NEAS can enhance the energy absorption performance of the steel tube. Based on this technique, systems of higher energy absorption capacities or of smaller sizes/masses can be designed.

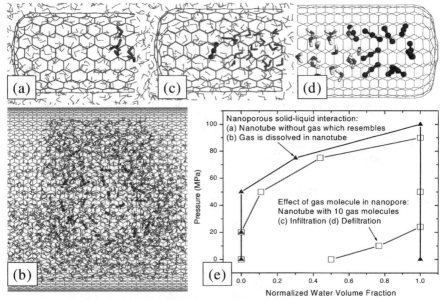

Fig.6 MD simulations of (a) infiltration of water (red) molecules into a vacuum small pore; (b) infiltration of water into a large pore containing CO_2 (blue) molecules; (c) infiltration of water into a small pore with 1 CO_2 molecule; (d) reduced gas solubility in small nanopore; (e) distinct infiltration/defiltration behaviors in large and small pores due to gas effect.

In order to gain more insights into the buckling mechanism, finite element (FE) simulations were carried out for both the empty and the NEAS filled steel tubes by using ABAQUS. The geometry, the Young's modulus (190 GPa), the yield stress (520 MPa), and the strain hardening exponent (0.4) of the tube were the same as that in the experiment. A rigid piston was placed at the top surface with a displacement-controlled motion, and the bottom of the tube was clamped rigidly. To simulate the behavior of the NFM liquid, a phenomenological approach was taken: the tube was first filled with a liquid with the bulk modulus of 2.1GPa, the same as that of water; upon loading, the pressure of the liquid quickly increased to 18 MPa, at which infiltration and the cell-wall buckling started. Using the pressure-volume curve of NEAS measured earlier [3], during the numerical simulation, the effective bulk modulus of the liquid filler was adjusted to decrease as a function of the cell volume. Finally, from both the force and the displacement acting on the piston, the relationship between the nominal stress and strain was computed, as shown in Fig.5.

It can be seen that the simulation has qualitatively captured the buckling initiation condition of the pressurized cell. The calculated critical nominal stress at the onset of cell-wall bucking is about 40 MPa, in good agreement with the experimental data. In terms of the buckled shape, the NEAS filled cell wrinkles and folds outward, and thus the wrinkle pattern is quite regular; whereas the empty cell may fold inward, and the pattern is relatively jerky, fitting well with the experimental observations (see Fig.5). In the empty cell, the nominal pressure largely decreases after buckling, and then the load oscillates as more wrinkles are formed and folded. In the NEAS filled cell, upon buckling, the nominal pressure is slightly reduced first due to the outward bulge of tube wall, and then the pressure remains at a plateau as the infiltration takes place in nanopores. Note that, prior to the buckling, the extended yielding of tube wall occurs along the radius direction. If this effect were ignored, the calculated buckling stress and wrinkle size would be much larger than the experimental data. Due to the over-simplification of the boundary conditions and the material properties, the relaxation valley "CD" as well as the second increase in load in Fig.3 cannot be simulated.

SCIENCE OF NEAS

In order to gain a deep insight into behaviors of pressurized nanofluids, pioneering molecular dynamics (MD) simulations are carried out on the liquid-gas-nanopore interactions [4]. Hydrophobic carbon nanotubes (CNTs) surrounded by water molecules are employed as a close analog to the symmetric compression of liquid/gas phases from both ends of long nanopores. In a large nanopore, the gas molecules are quickly dissolved in the infiltrated water, leading to a nanoenvironment that is dominated by solid-liquid interactions (Fig.6b) and analogous to an initially vacuum nanopore without gas molecules (Fig.6a). Whereas, when the nanopore size becomes smaller than about 2nm, the gas molecules cannot diffuse through the tiny space between the infiltrated water molecules and the CNT wall, thus they become undissolveable and form a gas cluster (Fig.6d).

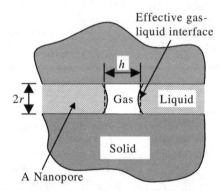

Fig.7 A schematic diagram of the gas-liquid phase transformation in a nanopore. The expansion of the gas phase leads to the "outflow".

The distinct gas solubility leads to fundamentally different infiltration and defiltration behaviors (Fig.6e): when the gas molecules are absent, a large pressure is needed to trigger the infiltration of water molecules into the energetically unfavorable nanoenvironment (Fig.6a). Once the CNT is fully occupied, after the pressure is reduced the water molecules prefer to stay inside the tube, in part because of the van der Waals attraction from solid molecules and in part because they are of a lower energy state after losing two hydrogen bonds to enter the pore. Thus, "nonoutflow" behavior is predicted for either small pores without gas molecule or for large pores where the gas molecules are dissolved by the infiltrated water (Fig.6b). By contrast, in a small CNT initially occupied by CO_2 molecules, due to the attraction force exerted by the CO_2, water molecules can infiltrate the pore at a lower threshold (Fig.6c). At peak pressure, a gas cluster is formed at the end of the nanotube (Fig.6d). Subsequently, when the system pressure is reduced, the gas cluster repels quite a few H_2O out of the CNT, leading to an "outflow" phenomenon that is exclusive to small nanopores with gas molecules. The importance of gas phase, which is systematically investigated for the first time, must be taken into consideration for nanofluidic studies, since in reality the presence of gas molecules in nanopores is almost inevitable, which has important effect on the reusability of NEAS.

Inspired by the MD simulation, a simple thermodynamics analysis is performed. In a liquid-filled nanopore (Fig.7), the nucleation of a gas "nanobubble" would increase the gas-liquid interfacial energy and reduce the liquid-solid interfacial energy; when the difference is balanced with the gas nucleation energy and the external work, a critical pore radius is derived, $r_{cr} = 2 \Delta\gamma/ [(p+\Delta\mu)+2\gamma_g/ h]$, where h is the size of the gas nucleus, $\Delta\mu$ is the specific nucleation energy, and γ_g is the effective gas-liquid interfacial tension. For a given h, when $r < r_{cr}$, the gas-phase nucleus is stable and can grow spontaneously, which would eventually lead to the "outflow"; otherwise the liquid would remain in the nanopore and the "nonoutflow" occurs. The value of r_{cr} is estimated to be ~1.5nm. In this framework, the infiltration of liquid into a nanopore is associated with the shrinkage of the gas phase; and when the gas phase grow and eventually occupy the entire porous space, equivalently speaking, liquid defiltration occurs.

The fundamental role of gas molecules is verified in the parallel experiment on a NEAS, [4] which consists of hydrophobic mesoporous silica particles with pore size ranging from $r = 1$-20nm, and specific surface area $A \approx 250 m^2/g$. The particle size is $D \approx 15-35\mu m$. The nanoporous particles are immersed in water and sealed in a steel container; the water pressure (p) is then increased (Fig.8). In a system with a large pore size ($r>10nm$), with increasing p, initially, as shown in Fig.9a, from point "O" to "A" the sorption isotherm is linear, with the effective bulk modulus (2GPa) close to that of water. As the pressure exceeds the infiltration pressure $p_{in} = 6$ MPa, the liquid overcomes the capillary effect and started to enter the nanopores, forming the plateau region from "A" to "B". When most of the pores were filled, the isotherm slope increases rapidly and the system responses linearly again ("B" to "C"). As the pressure is reduced, the confined liquid does not come out of the nanopores, as predicted by the MD simulations, leading to a pronounced hysteresis. The area enclosed by the hysteretic loop represents the energy absorption efficiency, E^*. After the first cycle, due to the "nonoutflow" of the liquid, the nanopores remain being occupied and the absorption ability is lost.

By contrast, in Fig.9b when the pore size is smaller than about 2nm (with other conditions being similar), the "outflow" behavior is observed at unloading and the system is fully recoverable. A remarkable phenomenon is, if the system in Fig.9b was hold at peak pressure for 12 hours, a large number of gas bubbles 0.1-1 mm in size were detected, and then when the pressure was lowered, "nonoutflow" occurred and consequently from the second loading cycle the energy absorption capacity was largely reduced (Fig.9c). Clearly, during the high-pressure resting, gas molecules that were initially trapped in the nanopores diffused out. As a result, the gas phase nucleation in nanopores became difficult and the confined liquid could not defiltrate. In another experiment on a hydrophobic silicalite of an even smaller pore size ($r\sim 0.6nm$), the liquid defiltration was much easier and the sorption isotherm was nearly non-hysteric (Fig.9d).

All observations qualitatively agreed with the MD simulation, which provides critical guidelines for adjusting the NEAS performance through simulation-based and object-oriented design: for example, small pore size is desired if the system is to subjected to cyclic loadings, e.g.

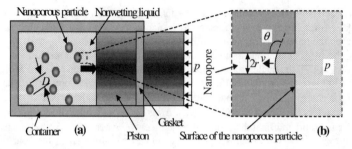

Fig.8 Schematic diagrams of (a) a system consisting of nanoporous particles immersed in a nonwetting liquid; and (b) the forced "flow" into a nanopore under external pressure.

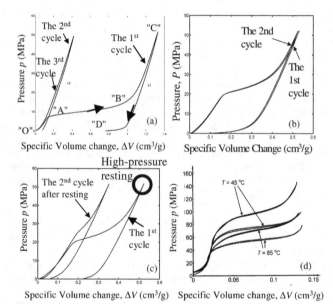

Fig.9 The sorption isotherms: (a) the NEAS of large r for one-time use; (b) the NEAS of small r for cyclic use; (c) after removing gas molecules in (b) the system is recoverable; (d) the non-hysteric sorption isotherm of zeolite based systems at different temperatures (solid line: pure water based; dash line: electrolyte solution based)

for damping stages (Fig.9c) and actuation systems (Fig.9d); and large pore sizes are ideal for one-time use, e.g. car bumpers and liquid armors (Fig.9a).

CONCLUSION

The energy absorption efficiency of the NEAS discussed above is much higher than that of conventional materials, demonstrating the great potential of applying nanoporous materials in protection devices. It provides an illustrative example of the proposed system, where the effects of various material/system parameters (effects of liquid/gas species, pore radius, pore surface property, temperature, internal friction, loading rate, etc.) can be explored by multi-scale simulations, and the science of nanofluidics can be used to guide the experiment and to optimize the manufacturing and performance of the active nanoporous systems.

Acknowledgements

The experimental work was supported by The Army Research Office under Grant No. W911NF-05-1-0288. The numerical simulation was supported by the National Science Foundation under Grant No. NSF-CMS0407743 and CMMI-0643732.

References

1 X. Kong, Y. Qiao, *Phil. Mag. Lett.* **85**, 331 (2005)
2 X. Kong, Y. Qiao, *Appl. Phys. Lett.* **86**, 151919 (2005)
3 X. Chen, F. Surani, X. Kong, V. Punyamurtula, Y. Qiao, *Appl. Phys. Lett.* **89**, 241918 (2006)
4 Y. Qiao, G. Cao, X. Chen, *J. Am. Chem. Soc.* **129**, 2355 (2007)

Mater. Res. Soc. Symp. Proc. Vol. 1041 © 2008 Materials Research Society   1041-R02-06

Novel Organometallic Fullerene Complexes for Vehicular Hydrogen Storage

Erin Whitney, Anne C. Dillon, Calvin Curtis, Chaiwat Engtrakul, Kevin O'Neill, Mark Davis, Lin Simpson, Kim Jones, Yufeng Zhao, Yong-Hyun Kim, Shengbai Zhang, and Philip Parilla
National Renewable Energy Laboratory, Golden, CO, 80401

ABSTRACT

Experimental wet chemical approaches have been demonstrated in the synthesis of a new chainlike $(C_{60}-Fe-C_{60}-Fe)_n$ complex. This structure has been proposed based on ^{13}C solid-state nuclear magnetic resonance, electron paramagnetic resonance, high-resolution transmission electron microscopy, energy-dispersive spectroscopy, and X-ray diffraction. Furthermore, this structure has been shown to have unique binding sites for dihydrogen molecules with the technique of temperature-programmed desorption. The new adsorption sites have binding energies that are stronger than that observed for hydrogen physisorbed on planar graphite, but significantly weaker than a chemical C-H bond. Volumetric measurements at 77 K and 2 bar show a hydrogen adsorption capacity of 0.5 wt%. Interestingly, the BET surface area is ~31 m^2/g after degassing, which is approximately an order of magnitude less than expected given the measured experimental hydrogen capacity. Nitrogen and hydrogen isotherms performed at 75 K also show a marked selectivity for hydrogen over nitrogen for this complex, indicating hidden surface area for hydrogen adsorption.

INTRODUCTION

A hydrogen-based economy offers the pollution-free promise of using entirely renewable resources.[1] For example, hydrogen can be generated through the electrolysis of water using electricity derived from wind power, photovoltaics, or thermo-chemical processing of biomass. Once produced, hydrogen can then be used in fuel cells that convert hydrogen and oxygen back into water and produce electricity in the process. Hydrogen can also be combusted in an engine to generate mechanical energy or even burned to produce heat. Regardless of the scenario, water is produced in a virtually pollution-free cycle.[1]

However, one of the biggest challenges facing a future hydrogen economy is that of onboard vehicular hydrogen storage. Hydrogen is a nonpolarizable gas, making reversible solid state hydrogen storage a difficult challenge. Furthermore, neither compression of H_2 to 10,000 p.s.i. or liquid hydrogen will satisfy all of the United States Department of Energy's 2015 targets for hydrogen storage of 9 wt% or 81 kg H_2/m^3.[2,3] Thus, in recent years, research has focused on novel carbon-based nanostructured materials, among others, as candidates for vehicular storage.[4,5] Carbon is promising because it is a light element and thus a step towards the goal of lightweight hydrogen storage for transportation.

Also inherent in the goal of hydrogen storage are the issues of near-room temperature operation at reasonable pressures. For an adsorption system, these challenges dictate a moderate binding energy for managing the heat load during refueling. Furthermore, the entire process must be completely reversible.[4] Although not typically appreciated, the adsorption energies for hydrogen bound to carbon surfaces are, in general, quite weak or quite strong. Non-dissociative physisorption, due purely to van der Waals interactions, involves a binding energy of only ~4 kJ/mol, whereas a C-H chemical bond is typically close to 400 kJ/mol. The desired binding energy range for reversible vehicular storage (~10-40 kJ/mol) therefore dictates that hydrogen be stabilized in an atypical fashion.

Hydrogen adsorption using carbon-based nanostructured materials has previously been explored on singled-wall nanotube (SWNT) structures,[6] with a binding energy of ~19 kJ/mol, as well as multi-wall nanotube (MWNT) structures grown with an iron catalyst (~50 kJ/mol).[5] In particular, the observation of enhanced hydrogen storage capacities in MWNT structures with small amounts of iron has fueled investigation of other potential metal-containing storage structures. While theory has predicted that isolated transition metal atoms can complex with up to six dihydrogen ligands via a Kubas interaction,[7-11] these metal atoms are predicted to form a bulk material upon removal of the hydrogen. To overcome this challenge and yet also harness the large storage potential of transition metal atoms, buckyballs have been proposed as stabilizing ligands because of their symmetric arrangement of cyclopentadiene rings, which have been shown to complex with transition metals through Dewar coordination,[12] but would otherwise polymerize without the presence of the buckyball matrix.

Fullerenes, or "buckyballs," are closed structures comprised of unsaturated carbon atoms arranged in 5- and 6-membered rings, providing a number of possible bonding modes for metal coordination. The most famous fullerene, C_{60}, was discovered by Kroto et al in 1985.[13] Calculations have shown that an iron atom can form an organometallic complex with a C_{36} fullerene, sharing charge with only four carbon atoms of a bent five-membered ring. Three molecular H_2 ligands then coordinate with the iron atom with a binding energy of ~43 kJ/mol. Notably, stable transition metal-coated buckyballs (Ti, V, Nb, Ta) have been recently synthesized.[14]

In an optimized fullerene-based transition metal complex, scandium has been predicted to complex with the twelve five-membered carbon rings of a fullerene, sharing charge with all of the carbon atoms in the pentagon ($\eta 5$ coordination) through Dewar coordination. For example, a $C_{60}[ScH_2(H_2)_4]_{12}$ organometallic fullerene complex (OFC) is predicted to be a minimum energy structure with ~7.0 wt% reversible hydrogen capacity.[15] Doping this OFC with boron results in a $C_{48}B[ScH_2(H_2)_4]_{12}$ OFC with a reversible hydrogen capacity of ~9 wt%. The complexes are arranged symmetrically on a buckyball in a minimum-energy structure, and the hydrogen is stored reversibly with a binding energy of ~30 kJ/mol, ideal for vehicular applications.

Although stable transition metal-coated buckyballs have been synthesized, the synthesis of these abovementioned $\eta 5$ complexes is unprecedented and many hurdles must be overcome. For example, the chemistry of C_{60} is generally olefinic (i.e., $\eta 2$ coordination, in which the metal is coordinated to the fullerene through two carbon atoms contributing two electrons to the bonding).[16-21] Thus, the synthesis of the predicted fullerene-metal-H_2 complexes, where the metal is coordinated to five carbon atoms, is not expected to be easy. In fact, η 5-C_{60} coordination has only been achieved through wet chemical methods by isolating the carbon atoms of a C_{60} pentagon through five-fold addition of alkyl-groups to neighboring carbon and protonation of one of the pentagon carbons.[22] Thus, the synthesis of the predicted fullerene-metal-H_2 complexes, where the metal is coordinated to five carbon atoms, is not expected to be easy.

However, the calculations described above, together with others that have recently appeared, indicate that non-olefinic metallofullerenes[23] and metal-doped nanostructures[24] are stable. Because the synthesis of these complexes is relatively unexplored and there are no guiding precedents, it has been necessary to discover the bonding preferences of the fullerene system and to open new synthetic pathways to the desired complexes. Here we describe the characterization of an iron atom, complexed with C_{60} ligands in a chainlike structure. New adsorption sites for dihydrogen molecules on carbon surfaces are clearly demonstrated.

EXPERIMENTAL PROCEDURES

To make the reactive fulleride compound K_6C_{60}, fullerenes and a slight excess of potassium were sealed in a glass tube under vacuum and heated for approximately four days at 250 °C. Both solid-state ^{13}C NMR and Raman spectroscopy were employed to determine that the K_6C_{60} compound was in fact synthesized. The K_6C_{60} product was then reacted in an inert atmosphere with cyclopentadienyl-iron-dicarbonyl-iodide (CpFe(CO)$_2$I) in tetrahydrofuran (THF) to form the complex. The recovered product was dried in an inert atmosphere. Manipulations of air-sensitive materials were carried out in a glove box or using standard Schlenk techniques. THF was distilled just prior to use from sodium benzophenone ketyl. C_{60} was obtained from Aldrich, and CpFe(CO)$_2$I was obtained from Strew.

It was difficult to dissolve the new complex in any organic solvent, making solid-state nuclear magnetic resonance (NMR) necessary. For these studies, a BRUKER AVANCE 200 spectrometer operating at 200 MHz was employed. Solid-state ^{13}C NMR spectroscopy under fast magic angle spinning (MAS) was required to obtain high-resolution spectra of the complex.[25] Transmission electron microscopy (TEM) was performed on a F-20 UT Transmission Electron Microscope with dry samples on a grid. X-ray diffraction (XRD) was performed on a Scintag PTS 4-circle goniometer (Bragg-Brentano geometry) using Cu Kα radiation (0.15406 nm) generated at 45 kV and 36 mA and detected with a liquid-nitrogen-cooled solid-state germanium detector. The source slits were 4 mm and 2 mm at 290 mm goniometer radius and the detector slits were 1.0 mm and 0.5 mm at the same radius. The sample powder was mounted onto a low-x-ray-background quartz substrate using diluted Duco cement. (The sample mount is vertical so the glue is necessary; the diluted glue adds almost no background signal and is amorphous.) The scan rate was 0.12 degrees/min. (25 seconds/step) from 5 to 125 degrees two theta in 0.05 degree steps (total time = 15.3 hours).

The new hydrogen binding sites were examined with temperature-programmed desorption (TPD) spectroscopy. The sample was first dosed with 500 Torr hydrogen for ~5 minutes after pumping at a pressure of approximately $5x10^{-8}$ Torr overnight. The sample was then cooled to liquid nitrogen temperatures and systematically degassed to temperatures up to 250 °C. For the TPD technique, samples weighing between 1-10 mg are placed in a packet formed from 25 μm thick platinum foil and mounted at the bottom of a liquid nitrogen cooled cryostat. The packet is resistively heated with a programmable power supply, and the sample temperature is measured with a thin thermocouple spot-welded to the platinum packet. A mass spectrometer measures desorbing species and insures that only hydrogen is observed during desorption. The TPD instrument is calibrated by thermally decomposing known amounts of TiH_2. The amount of evolved hydrogen is linear with the weight of decomposed TiH_2. The TPD system is also calibrated by H_2 desorption from Pd that is charged *in situ* to a literature predicted capacity. Finally a calibrated H_2 flow is employed as a further check of the calibration standards. All of the methods have been confirmed with an in-house volumetric technique within ±3 %. A control TPD experiment, using C_{60}, was also done as a comparison. Additionally, using the same dosing and cooling techniques, the new Fe-C_{60} compound was heated at a variety of different rates in order to extract a desorption activation energy E_d.[6] Total H_2 capacity measurements for the bucky dumbbell complex were obtained at 77K and 2 bar with single point measurements in a volumetric apparatus.

RESULTS AND DISCUSSION

Structural Characterization

Figure 1 displays the ^{13}C solid-state NMR spectrum characteristic of C_{60}. The sharp peak with a chemical shift of 143.7 ppm is consistent with C_{60}, and the broad feature at approximately 110 ppm is simply due to the Teflon cap sealing the NMR rotor.

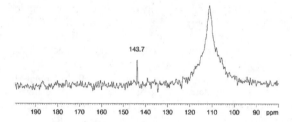

Figure 1: ^{13}C NMR spectrum of C_{60} fullerenes.

In comparison, Figure 2 displays the ^{13}C solid-state NMR spectrum of the final recovered product from synthesis reactions designed to unfold new organometallic chemistry for C_{60}. Here the sharp peak at 143.7 ppm is again attributed to residual unreacted C_{60}. The broad peak shifted to higher ppm is consistent with C_{60} coordinated with an iron atom. An inset in the figure of these two peaks is also provided for clarity.\

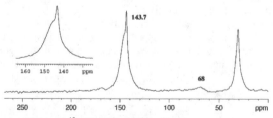

Figure 2: ^{13}C NMR spectrum of Fe-C_{60}.

In order to better elucidate the precise structure of the Fe-C_{60} compound, high-resolution transmission electron microscopy (HRTEM) and energy dispersive x-ray (EDX) spectroscopy were performed with a field emission microscope. As shown in Fig. 3, the study revealed areas in the sample, highlighted by pink circles, that were consistent with small quantities of oxidized iron in the phase Fe_2O_3. The sample was exposed to air as the complex was not found to be air-sensitive during hydrogen adsorption studies. However, it was not surprising that small amounts of residual iron were oxidized immediately. More importantly, regions of C_{60} molecules, circled in yellow, containing ~1-1.5 at% iron were also observed. The fact that such low levels of iron were stable against oxidation suggests that the iron is complexed to the C_{60} molecules and is consistent with the formation of C_{60}-Fe-C_{60}-Fe-C_{60}-chain structures of a yet-undetermined length. From the HRTEM image in Fig. 3, some ordering in the C_{60} chain-like structures may be detected. Also, no large metal clusters were observed with extensive HRTEM analyses.

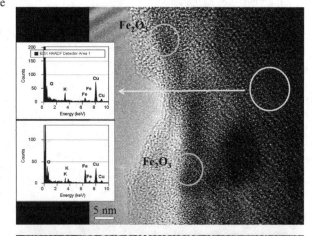

Figure 3: HRTEM of air-exposed Fe-C_{60} sample.

HRTEM analysis was repeated after washing the sample with dilute acid to remove Fe_2O_3. As shown in Figure 4, ordered chains emerge very distinctively, and an electron diffraction pattern was also obtained. Iron was still detected at ~0.5-1 at% with small spot EDX, again consistent with C_{60}-Fe-C_{60}-Fe-C_{60}-chain structures.

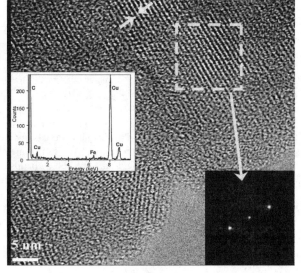

Figure 4: HRTEM of acid-washed Fe-C_{60} sample.

Since some ordering was detected in the HRTEM studies, XRD measurements were therefore performed on the powdered $Fe(C_{60})$ sample. Fig. 5(a) displays the XRD pattern for the $Fe(C_{60})$ sample. Several very broad features are observed that could be consistent with either disordered C_{60} or the disordered fulleride (used as a reactive precursor in the initial reaction). However, these broad features could also be consistent with disordered C_{60}-Fe-C_{60}-Fe-C_{60}-chain structures. Furthermore, the sharper feature occurring at low angle is consistent with a crystalline d-spacing of ~ 13.3 Å. This d-spacing is not consistent with the FCC packing of either C_{60} or the fulleride and suggests that a new packing of C_{60} is observed and may be attributed to $Fe(C_{60})$ chain-like structures. The broad features in the XRD pattern of Fig. 5(a) are also similar to features previously reported for carbon single wall

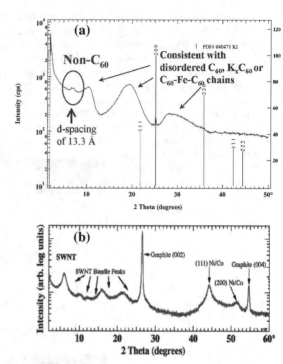

Figure 5: XRD spectra of (a) Fe-C_{60} and (b) SWNTs

nanotubes (SWNTs), as shown in Fig. 5(b). In the case of SWNTs, features are detected with XRD because the nanotubes pack into crystalline bundles. The features are broad, however, because the bundles are highly disorderd; i.e. there is slippage along the axis of the tubes such that the graphitic structures are not perfectly aligned.

Collectively this data suggests that the C_{60}-Fe-C_{60}-Fe-C_{60}-chain structures pack in loosely ordered bundles similar to SWNTs. It is possible that the 13.3 Å d-spacing represents the interstitial spacing between the chains. Fig. 6 displays a cartoon representation of Fe-C_{60} chains. (Note that based on the XRD in the actual materials, the degree of alignment is not expected to be this high.) It has been proposed that SWNTs that are atomically doped with metals are still a promising hydrogen storage material. The FeC_{60} chain structures are very similar to SWNTs doped with atomic metal. The focus of future work will be the production of these materials at higher yield so that they may be more readily purified, as well as methods to increase their alignment. An optimized porous framework for hydrogen storage at ~77 K with a moderate over-pressure may then be realized.

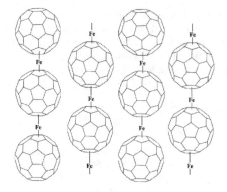

Figure 6: Cartoon representation of Fe-C_{60} chains.

New Hydrogen Adsorption Sites

Figure 7 displays temperature-programmed desorption spectra where an Fe-C_{60} sample was exposed to hydrogen at room temperature after pumping at a pressure of approximately 5×10^{-8} Torr overnight and then after sequentially degassing the sample to 100, 200 and 250 °C. The lower temperature peak is centered at approximately 100 °C. This is slightly above the peak desorption temperature that is generally observed for hydrogen adsorbed on carbon surfaces. (In fact, it is generally the case that the true desorption peak for physisorbed hydrogen is not obtained due to the inability to cool the sample below -140 °C while exposing the sample to H_2 at 500 Torr.) Perhaps more interesting, however, is the appearance of the peak centered at approximately -50 °C as the sample is degassed to 250°C. (The sample was not degassed above this temperature as organometallic complexes are known to decompose at temperatures ≥ 300 °C.) The appearance of this new peak shows that hydrogen is stabilized at a temperature significantly above that expected for physisorption. However, the temperature is significantly low enough that the probability of the formation of C-H bonds is essentially zero.

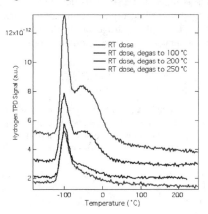

Figure 7: TPD spectra of Fe-C_{60} following a room temperature exposure to hydrogen, (500 Torr, 5 min.) after pumping at a pressure of ~ 5×10^{-8} Torr overnight and then after sequentially degassing the sample to 100, 200 and 250 °C.

The appearance of this new peak has also been compared with H_2 adsorption on C_{60} alone, as shown in Figure 8. Again, temperature-programmed desorption reveals H_2 adsorption sites following exposure to hydrogen at 500 torr for five minutes and then cooling approximately to -180 °C. Both samples were degassed to 250 °C prior to H_2 exposure. As the figure shows, the pure fullerene exhibits almost no hydrogen adsorption under these conditions. Thus, new adsorption sites for dihydrogen molecules have been revealed on C_{60}-Fe-C_{60}.

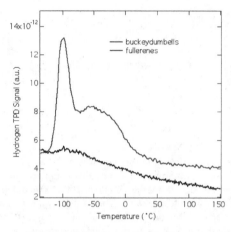

Figure 8: Comparison of H_2 adsorption on C_{60} and C_{60}-Fe-C_{60}. Sample was dosed at 500 torr for 5 minutes before cooling to -180 °C. Both samples were degassed to 250 °C prior to exposure.

For the unique adsorption site occurring at a higher temperature in the new complex, the exact binding energy (or desorption activation energy, E_d) can be determined by measuring the desorption peak temperature at different heating rates. Figure 9 shows a subsequent plot, described by $\ln T_m = E_d/RT_m$. Each point is derived from a different heating rate, which shifts the peak temperature of desorption. The slope of the line indicates an enhanced binding energy of ~6.2 kJ/mol, near the desired binding energy range for reversible onboard vehicular hydrogen storage. Since the binding energy was found to be only slightly enhanced over that observed for physisorption, non-dissociative adsorption is assumed and E_d is equivalent to the binding energy.[26] Complimenting the temperature programmed desorption studies, volumetric analyses conducted at 77K and 3 bar showed that the Fe-C_{60} had a hydrogen adsorption capacity of ~0.5 wt %. Furthermore, essentially zero hydrogen uptake was observed on pure fullerenes under the same conditions.

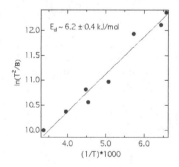

Figure 9: A plot of desorption activation energy indicated an enhanced binding energy of ~6.2 kJ/mol for the desorption peak centered at -50 °C, as shown in Fig. 7.

BET surface areas were calculated for the Fe-C_{60} complex, as synthesized and after degassing. For the as-synthesized compound, the surface area was measured twice, yielding results of 8.7 and 9.8 m^2/g. The gravimetric hydrogen capacity was 0.004 wt% at room temperature and 0.26 wt% at 75 K. After degassing the sample to 285 °C, the surface area was 31.1 m^2/g, with gravimetric hydrogen capacities of 0.004 wt% at room temperature and 0.5 wt% at 75 K. These results, especially for the degassed sample, possibly suggest a mechanism other than simple physisorption and also violate Chahine's rule. If there is not a unique mechanism, then there are surface sites which are not accessible to the N_2 used in these surface area measurements.

In light of these surprising results, N_2 and H_2 isotherms were run on the Fe- C_{60} complex at 75 K. As displayed in Figure 10, the H_2 isotherm reveals a dramatically faster uptake than that of N_2, suggesting that pore size may play a role in the hydrogen capacity of this compound. To further illuminate the debate between surface area effects and bonding effects, a CO_2 isotherm at 0 K was also performed (see Figure 11) and yielded a surface area of ~170 m^2/g. The rationale for using CO_2 is its higher kinetic energy and diffusion coefficient. Finally, in addition to the previously mentioned H_2 isotherm at 75 K, another H_2 isotherm at 80 K was conducted in order to estimate an enthalpy of adsorption of ~5 kJ/mol.

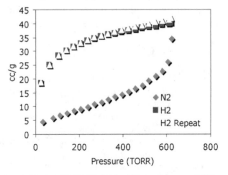

Figure 10: Fe-C_{60} H_2 and N_2 isotherms at 75 K

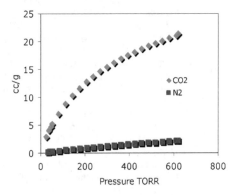

Figure 11: F-C_{60} CO_2 and N_2 isotherms at 0 °C

CONCLUSIONS

The formation of a new organometallic C_{60}-Fe-C_{60}-Fe chainlike structure has been demonstrated, and a unique hydrogen adsorption site on this complex has also been observed that is not anticipated for physisoprtion on a carbon surface. Furthermore, since the peak desorption temperature is -50 °C, dihydrogen species are most likely stabilized on the surface. The binding energy of this dihydrogen species will also likely be within the desired range for onboard vehicular hydrogen storage.

This new complex has been characterized with solid state NMR spectroscopy, XRD, HRTEM, and temperature-programmed hydrogen desorption. Analysis of the iron-fullerene complex indicates the formation of C_{60}-Fe-C_{60}-Fe-C_{60} chain structures of an undetermined length, with a reversible hydrogen capacity of ~0.5 wt% at 77 K and a hydrogen overpressure of 2 bar. Interestingly, the BET surface area is ~31 m^2/g after degassing, which is an order of magnitude less than expected given the measured experimental hydrogen capacity. Nitrogen and hydrogen isotherms performed at 75 K also show a marked selectivity for hydrogen over nitrogen for this complex, indicating hidden surface area for hydrogen adsorption. These results suggest that synthesis of organometallic fullerene complexes should be further explored for vehicular hydrogen storage applications.

REFERENCES

[1]   H.-H. Rogner, Int. J. Hydrogen Energy **23**, 833 (1998).

[2]   *http://www.eere.energy.gov/hydrogenandfuelcells/mypp/*.

[3]   *http://www.sc.doe.gov/bes/hydrogen.pdf*.

[4]   A. C. Dillon and M. J. Heben, Appl. Phys. A **72**, 133-142 (2001).

[5]   A. C. Dillon, J. L. Blackburn, P. A. Parilla, Y. Zhao, Y.-H. Kim, S. B. Zhang, A. H. Mahan, J. L. Alleman, K. M. Jones, K. E. H. Gilbert, and M. J. Hebern, in *Discovering the Mechanism of H_2 Adsorption on Aromatic Carbon Nanostructures to Develop Adsorbents for Vehicular Applications*, Boston, Massachusetts, 2004 (Materials Research Society), p. 117-124.

[6]   A. C. Dillon, K. M. Jones, T. A. Bekkedahl, C. H. Kiang, D. S. Bethune, and M. J. Heben, Nature **386**, 377-379 (1997).

[7]   G. J. Kubas, R. R. Ryan, B. I. Swanson, P. J. Vergamini, and H. J. Wasserman, J. Am. Chem. Soc. **106**, 451-452 (1984).

[8]   G. J. Kubas, J. Organometall. Chem. **635**, 37-68 (2001).

[9]   T. Le-Husebo and C. M. Jensen, Inorg. Chem. **32**, 3797-3798 (1993).

[10]   J. Niu, K. Rao, and P. Jena, Phys. Rev. Lett. **68**, 2277-2280 (1992).

[11]   F. Maseras and A. Lledos, Chem. Rev. **100**, 601-636 (2000).

[12]   D. Michael and P. Mingos, J. Organometall. Chem. **635**, 1 (2001).

[13]   H. W. Kroto, J. R. Heath, S. C. O'Brien, R. F. Curl, and R. E. Smalley, Nature **318**, 162-163 (1985).

[14]   F. Tast, N. Malinowski, S. Frank, M. Heinebrodt, I. M. L. Billas, and T. P. Martin, Phys. Rev. Lett. **77**, 3529-3532 (1996).

[15]   Y. Zhao, Y.-H. Kim, A. C. Dillon, M. J. Heben, and S. B. Zhang, Phys. Rev. Lett. **94**, 155504 (2005).

[16]   F. J. Brady, D. J. Cardin, and M. Domin, J. Organometall. Chem. **491**, 169-172 (1995).

[17]   P. J. Fagan, J. C. Calabrese, and B. Malone, Acc. Chem. Res. **25**, 134-142 (1992).

[18]   H.-F. Hsu, Y. Du, T. E. Albrecht-Schmitt, S. R. Wlson, and J. R. Shapley, Organometallics **17**, 1756-1761 (1998).

[19]   M. M. Olmstead, L. Hao, and A. L. Balch, J. Organometall. Chem. **578**, 85-90 (1998).

[20]   L.-C. Song, G.-A. Yu, F.-H. Su, and Q.-M. Hu, Organometallics **23**, 4192-4198 (2004).

[21]   D. M. Thompson, M. Bengough, and M. C. Baird, Organometallics **21**, 4762-4770 (2002).

[22]   M. Sawamura, M. Toganoh, Y. Kuninobu, S. Kato, and E. Nakamura, Chem. Lett. **29**, 270 (2000).

[23]   Q. Sun, Q. Wang, P. Jena, and Y. Kawazoe, J. Am. Chem. Soc. **127**, 14582-14583 (2005).

[24]   T. Yildirim and S. Ciraci, Phys. Rev. Lett. **94**, 175501 (2005).

[25]   C. Engtrakul, M. R. Davis, T. Gennett, A. C. Dillon, K. M. Jones, and M. J. Heben, J. Am. Chem. Soc. **127**, 17548-17555 (2005).

[26]   R. J. Madix, in *Chemistry and Physics of Solid Surfaces*, edited by R. Vanselov (CRC, Boca Raton, 1979), p. 63-72.

Novel Energy Storage Technologies

Mater. Res. Soc. Symp. Proc. Vol. 1041 © 2008 Materials Research Society

A Novel High Capacity, Environmental Benign Energy Storage System: Super-iron Boride Battery

Xingwen Yu[1], and Stuart Licht[2]
[1]Department of Chemical and Biological Engineering, The University of British Columbia, 2360 East Mall, Vancouver, V6T 1Z3, Canada
[2]Department of Chemistry, University of Massachusetts, 100 Morrissey Blvd, Boston, MA, 02125

ABSTRACT

High electrochemical capacity of alkaline boride anodes is presented. The alkaline anodes based on transition metal borides can deliver exceptionally high discharge capacity. Over 3800 mAh/g discharge capacity is obtained for the commercial available vanadium diboride (VB_2), much higher than the theoretical capacity of commonly used zinc metal (820 mAh/g) alkaline anode. Coupling with the super-iron cathodes, the novel Fe^{6+}/B^{2-} battery chemistry generates a matched electrochemical potential to the pervasive, conventional MnO_2-Zn battery, but sustains a much higher electrochemical capacity.

INTRODUCTION

Alkaline Zn-MnO_2 redox charge storage has been established for over a century, and is still playing the dominant share in primary alkaline battery market. However, this battery chemistry is increasingly limited in meeting the growing energy and power demands of contemporary optical, electromechanical, electronic, and medical consumer devices. Therefore, the search for new energy storage chemistry systems with higher capacity and energy density has been increasingly emphasized. A number of new materials, such as metal hydride and intercalation compounds have been successfully applied to the high performance Ni-MH and Li-ion batteries [1,2]. We introduced a new battery type, super-iron battery based on the high Fe(VI) cathodic charge storage in 1999 [3]. Followed the primary alkaline super-iron battery, recently, rechargable thin layer super iron cathode has been reported [4,5], and a high performance composite Fe(VI)/AgO composite cathode stabilized by a 1% zirconia coating has also been successfully developed [6]. In 2004 it was reported that metal borides could be used as anodic alkaline charge storage materials [7,8]. Representative transition metal borides include TiB_2 and VB_2 which can store several folds more charge than a zinc anode through multi-electron charge transfer: [7]

$$TiB_2 + 12OH^- \rightarrow Ti(\text{amorphous}) + 2BO_3^{3-} + 6H_2O + 6e^- \tag{1}$$
$$VB_2 + 20OH^- \rightarrow VO_4^{3-} + 2BO_3^{3-} + 10H_2O + 11e^- \tag{2}$$

However, one obstacle was evident towards implementation of this alkaline boride (MB_2, M = Ti or V) anodic chemistry. The electrochemical potential of the boride anodes was more positive than that of zinc. Therefore the voltage of a boride MnO_2 cell was low compared to the voltage of the pervasive Zn–MnO_2 battery. In our recent communication, we introduced a novel Fe^{6+}/B^{2-} battery chemistry in which the super-iron (Fe^{6+}) cathode provides the requisite additional electrochemical potential for the boride (B^{2-}) anode. Therefore, the Fe(VI)–MB_2

couple generates a similar potential to the Zn–MnO$_2$ battery. In addition, the obstacles of the boride anode decomposition are overcome by the applying a zirconia hydroxide-shuttle overlayer on the anode particle surface [9]. This paper is a continuous study of the Fe^{6+}/B^{2-} redox chemistry. More boride anodes and Fe(VI)/AgO composite cathodes are studied in super iron boride batteries.

EXPERIMENTAL

Batteries studied in this paper are prepared as 1 cm button cell configuration. The cells are prepared with a saturated KOH electrolyte. Preparation of the cathodes and anodes will be detailed in the Results & Discussion section. Cathode material used in this paper is the lab synthesized K$_2$FeO$_4$ (97-98% purity, according to our previous publication [10,11]). Anode materials TiB$_2$ (10 μm powder), VB$_2$ (325 mesh powder) TaB (325 mesh), TaB$_2$ (325 mesh), MgB$_2$ (325 mesh), CrB$_2$ (325 mesh), CoB$_2$ (325 mesh), Ni$_2$B (powder, 30 mesh) and LaB$_6$ (powder, 10 μm) are from Aldrich®. Conductive medium used in cathode and anode preparation is 1μ graphite (Leico Industries Inc.). The cells are discharged at a constant load (will be indicated in Results & Discussion section). Primary discharge is measured as the cell potential variation over time, and is recorded with LabView Acquisition on a PC, and cumulative discharge, as milliampere hours, determined by subsequent integration.

RESULTS AND DISCUSSION

Many transition metal borides have thermodynamic parameters and electronic conductivities similar to those of the corresponding transition metals. Therefore, from the electrochemical energy conversion viewpoint, transition metal borides may constitute a large class of promising electrochemically active materials for batteries [12,13]. In the study herein, various transition metal borides TaB, TaB$_2$, MgB$_2$, CrB$_2$, CoB$_2$, Ni$_2$B, LaB$_6$ are considered as the anodes for alkaline battery. Similarly as TiB$_2$ and VB$_2$ [7], electrochemical potentials of the borides are more positive than zinc metal, thus the cell voltage of MnO$_2$-boride batteries are lower than the conventional MnO$_2$-Zn battery. The alkaline thermodynamic potential of the 3-e$^-$ reduction of super iron cathodes Fe(VI → III) via equation (3), is approximately 250 mV higher than the one electron reduction of MnO$_2$ via equation (4), with potentials reported versus SHE (the standard H$_2$ electrode):

$$FeO_4^{2-} + 5/2H_2O + 3e^- \rightarrow 1/2Fe_2O_3 + 5OH^-; E = 0.60 \text{ V} \quad (3)$$
$$MnO_2 + 1/2H_2O + e^- \rightarrow 1/2Mn_2O_3 + OH^-; E = 0.35 \text{ V} \quad (4)$$

Coupling with super-iron cathode (K$_2$FeO$_4$), open circuit potentials of these super iron boride batteries are listed in Table I.

Table I. Open circuit potentials (OCP) of various boride anodes alkaline super-iron (K$_2$FeO$_4$) batteries. Electrolyte used is saturated KOH

Anode	TaB	TaB$_2$	CoB$_2$	MgB$_2$	CrB$_2$	Ni$_2$B	LaB$_6$	VB$_2$	TiB$_2$
OCP	1.65V	0.83V	1.22V	1.50V	1.53V	1.30V	1.16V	1.42V	1.55V

In addition to the previously studied TiB_2 and VB_2 [7,9], in Table I, only the first two (tantalum) boride salts exhibit a degree of anodic charge storage, each of the other borides did not exhibit significant primary discharge behavior due to their high solubilities in alkaline solution or their reaction with alkaline electrolyte. As previously reported [7,9], the alkaline oxidation of the TiB_2 anode produced amorphous titanium, and similarly we expect the reduction of TaB_2 can yield tantalum:

$$TaB_2 + 12OH^- \rightarrow Ta + 2BO_3^{3-} + 6H_2O + 6e^- \qquad (5)$$
$$TaB + 6OH^- \rightarrow Ta + BO_3^{3-} + 3H_2O + 3e^- \qquad (6)$$

According to the oxidation reaction of the half cell and the formula weight of TaB_2 (FW 202.57 g/mol) or TaB (FW 191.76 g/mol), intrinsic capacity of TaB_2 and TaB would accordingly be: 793.8 mAh/g and 419.3 mAh/g.

Discharge profiles of TaB and TaB_2 anode, K_2FeO_4 cathode button cells are shown in figure 1a. Button cells are prepared with excess cathode capacity and discharged at a low current to probe the anode's limits and characteristics. As seen in the figure, the TaB_2 exhibits an anodic storage capacity comparable to the widely used conventional alkaline zinc anode (820 mAh/g). Compared to the theoretical capacity (793.8 mAh/g and for TaB_2 and 419.3 mAh/g for TaB), 88% for TaB_2 and 93% for TaB of the coulombic efficiency were obtained for these 2 anodes.

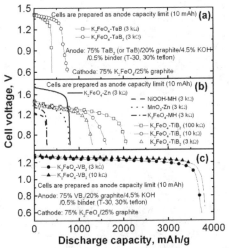

Figure 1. (a): Discharge profile of TaB_2(TaB)-K_2FeO_4 button cells. (b): Comparison of TiB_2 anode cells to cells with conventional Zn or MH alkaline anodes. As indicated the cells contain NiOOH, MnO_2 or K_2FeO_4 cathode. (c): Discharge of a comparable VB_2 anode/ K_2FeO_4 cathode alkaline cells.

The discharge of TiB_2, rather than TaB_2, to the amorphous metal product is comparable to a significantly higher gravimetric charge storage capacity due to the lighter weight of this metal (47.87 g titanium/mol compared to 180.95 g tantalum/mol), in accord with equation (1).

As seen in figure 1b, a significant advantage of the titanium boride anode is the higher capacity compared to the conventional alkaline zinc anode (820 mAh/g). The TiB_2 anode discharge is in excess of of 1300 mAh /g at moderate discharge rates (3kΩ load) and is in excess of 2000 mAh/g at low discharge rates (100 kΩ load). In accord with Eqn (1), and a formula weight, W =69.5 g mol^{-1}, TiB_2, has a net intrinsic 6 electron anodic capacity of 6F/W = 2314 mAh/g (F= the faraday constant).

While the TiB_2 exhibits greater intrinsic capacity than the TaB_2, a vanadium, compared to a titanium, diboride salt can yield even higher alkaline anodic capacity. Unlike TiB_2, the alkaline VB_2 anode undergoes an oxidation of both the boron and the tetravalent transition metal ion, with a net 11 electron process. In accord with Eq. (2), VB_2, will have an intrinsic anodic capacity of 11F/(W=72.6 g mol^{-1}) = 4060 mAh/g, rivaling the high anodic capacity of lithium (3860 mAh/g). As evident in figure 1c, this substantial capacity of VB_2 is experimentally realized (3800 mAh/g) in the discharge of the alkaline super-iron VB_2 cell. The VB_2 sustains more efficient higher rate discharge (and lower polarization loss) than the TiB_2 alkaline anode cell. As seen comparing the 3kΩ discharges in figure 1b and c, the TiB_2 discharge sustains 56% of the intrinsic capacity, whereas the VB_2 sustains 91% of the intrinsic capacity (which increases to 94% during a 10kΩ discharge).

We recently reported a novel K_2FeO_4/AgO composite cathode material for alkaline battery [6]. AgO additive facilitates the Fe(VI) charge transfer. Increasing a AgO additive in a composite K_2FeO_4 cathode remarkably improves its charge transfer at high discharge rate, as shown in figure 2a.

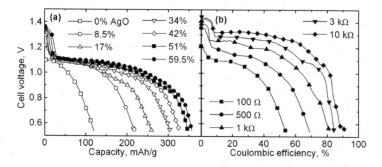

Figure 2. (a): High current discharge profiles of composite K_2FeO_4/AgO-TiB_2 button cells. The K_2FeO_4/AgO composite cathode is studied in cells with excess intrinsic anode capacity, in a 1 cm coin cell, discharged under 100 Ω constant load conditions. Fraction of AgO additive is from 0 to 59.5%. (b): Discharge voltage profiles at a range of discharge loads, of fixed composition composite K_2FeO_4/AgO-TiB_2 button cells.

With a large fraction of AgO additive, the composite cathode yields over four times capacity than that without AgO additive. This composite cathode is further studied by coupling with the exceptional high capacity TiB_2 anode. An effective composite cathode is prepared with 29.5 mg 76.5% K_2FeO_4, 8.5% AgO, 5% KOH and 10% graphite. Excess capacity TiB_2 anode is prepared as 75% TiB_2, 20% graphite, 4.5% KOH and 0.5% bender (T-30, 30% Teflon). Discharge profiles of this novel type of button cells at different constant loads are shown in figure 2b. The intrinsic capacity of the composite cathode is 408 mAh/g, based on 3-electron charge transfer of K_2FeO_4 (406mAh/g) and 2-electron charge transfer of AgO (432mAh/g) (90%*406+10%*432=408mAh/g). Depth of discharge of this composite cathode is relatively high at low current density discharge, and utilization of intrinsic capacity is over 80% (e.g. Discharged at 3000 Ω and 10000 Ω). However, only 45% of capacity is utilized at lower load, 100 Ω, discharge.

The range from maximum experimental (<160 mAh/g) to theoretical (2F per $Zn+2MnO_2$ = 222 mAh/g) charge storage capacities of the conventional alkaline $Zn-MnO_2$ cell are shown as vertical lines in figure 3. The theoretical capacities for the complete super-iron TiB_2 (6F per TiB_2 + $2K_2FeO_4$) and super-iron VB_2 (11F per VB_2 + $11/3K_2FeO_4$) are higher respectively at 345 and 369 mAh/g. The discharge of the complete super-iron boride redox chemistry is investigated in figure 3 using cells with balanced anode and cathode capacity (based on the intrinsic capacity of the anode and cathode components). As seen in figure 3, the super-iron titanium boride cell combined anode and cathode capacity is experimentally in excess of 250 mAh/g, and that of the super-iron VB_2 cell is over 310 mAh/g, which is two fold higher than that of the conventional alkaline battery chemistry (MnO_2/Zn). The super-iron boride chemistry exhibits significantly higher charge storage than conventional alkaline primary storage chemistry. A further optimization of both the boride and super-iron salt particle size should further enhance cell performance.

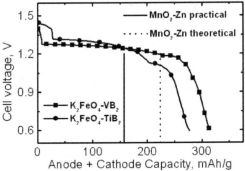

Figure 3. Capacity (anode+cathode) of the super-iron boride alkaline batteries to the conventional (MnO_2/Zn) alkaline battery. The super-iron boride cell contains either a TiB_2, or a VB_2 anode, as indicated in the figure. The cathode is 76.5% K_2FeO_4, 8.5% AgO, 5% KOH and 10% 1μm graphite.

CONCLUSIONS

A novel high capacity battery chemistry based on super iron cathode and transition metal boride anode is presented. This Fe^{6+}/B^{2-} redox chemistry generates a matched electrochemical potential to the pervasive, conventional MnO_2-Zn battery chemistry, but sustains a much higher electrochemical capacity. With the additive AgO to the super iron (K_2FeO_4) cathode, A K_2FeO_4/AgO composite cathode TiB_2 anode alkaline battery exhibits high-rate discharge performance.

REFERENCES

1. S. R. Ovshinsky, M. A. Fetcenko and J. Ross, *Science* **260**, 176 (1993).
2. G. M. Julien, *Mater. Sci. Eng. R* **40**, 47 (2003).
3. S. Licht, B. Wang and S. Ghosh, *Science* **285**, 1039 (1999).
4. S. Licht and Ran Tel-Vered, *Chem. Comm.* **6**, 628 (2004).
5. S. Licht and C. DeAlwis, *J. Phys. Chem. B* **110**, 12394 (2006).
6. S. Licht, X. Yu and D. Zheng, *Chem. Comm.* **41**, 4341 (2006).
7. H. X. Yang, Y. D. Wang, X. P. Ai and C. S. Cha, *Electrochem. Solid-State Lett.* **7**, A212 (2004).
8. Y. D. Wang, X. P. Ai, Y. L. Cao and H. X. Yang, *Electrochem. Comm.* **6**, 780 (2004).
9. S. Licht, X. Yu and D. Qu, *Chem. Comm.* **26**, 2753 (2007).
10. S. Licht, V. Naschitz, B. Liu, S. Ghosh, N. Halperin, L. Halperin and D. Rozen, *J. Power Sources* **99**, 7 (2001).
11. S. Licht, V. Naschitz, L. Halperin, N. Halperin, L. Lin, J. Chen, S. Ghosh and B. Liu, *J. Power Sources* **101**, 167 (2001).
12. S. Amendola, U.S. Patent No. 5 948 558 (1999).
13. S. Amendola, U.S. Patent No. 6 468 694 (2002).

Mater. Res. Soc. Symp. Proc. Vol. 1041 © 2008 Materials Research Society 1041-R03-02

Mechanical Effect on Oxygen Mobility in Yttria Stabilized Zirconia

Wakako Araki[1], and Tadaharu Adachi[2]
[1]Mechanical Sciences and Engineering, Tokyo Institute of Technology, 2-12-1-I6-5 O-okayama, Meguro-ku, Tokyo, 152-8552, Japan
[2]Mechanical Sciences and Engineering, Tokyo Institute of Technology, 2-12-1-I6-1 O-okayama, Meguro-ku, Tokyo, 152-8552, Japan

ABSTRACT

The mechanical effect on the oxygen ion mobility in zirconia stabilized with 8 mol% yttria was investigated in this study. A dynamic mechanical thermal analysis showed that the dynamic modulus decreased gradually with temperature while the mechanical loss had two peaks due to different relaxation mechanisms. From the comparison of activation energies between the ionic conductivity and the mechanical relaxation, the dominant factor for oxygen mobility was determined to be the migration of oxygen vacancies in the simple complexes. The result also illustrated the strong relationship between the modulus and the conductivity. An impedance analysis under mechanical tensile-loading conditions showed that the mechanical load improved the ionic conductivity by 6 % at maximum although the improvement was a temporary effect.

INTRODUCTION

Oxygen ionic conductive materials are promising materials as electrolytes for solid oxide fuel cells (SOFCs) and oxygen sensors and have been thoroughly investigated in recent years [1,2]. Zirconia and ceria doped with rare earths have been found to show an ionic conductivity due to oxygen vacancies. Especially, yittria-stabilized zirconia (YSZ) has been widely used as the oxygen ionic conductive material because of its high ionic conductivity as well as its excellent mechanical properties.

Much research have been done on the mechanical properties and behaviors of YSZ, such as its elastic modulus, bending strength [2], fracture toughness, superplasticity[3], and also stress-induced phase transformation [4]. The modulus and strength were experimentally shown to decrease with increasing temperature [5]. More recently, the mechanical loss, i.e. internal friction, of YSZ has been investigated and examined under various temperatures and frequencies [6-9]. It has been discussed that the mechanical loss can be attributed to migration of oxygen vacancies. Thus, the ionic conductivity could be strongly related to the mechanical viscoelastic relaxation behavior.

In addition to the relationship between the mechanical viscoelastic behavior and the conductivity, the effect of mechanical stress and strain on the conductivity is very interesting. Dielectric materials such as lead zirconate titanate are widely known to show the piezoelectricity due to intrinsic polarization caused by the mechanical stress. The electron mobility in the semiconductors such as doped silicon is known to be improved by mechanical strain, which cause the band structures. Few studies [10.11], however, have reported on the relationship between the mechanical stress or strain and the ionic conductivity. It has been found out that the plastic compressive deformation improved ionic conductivity in yttria-stabilized zirconia by means of dislocation mechanism [10]. On the other hand, other study have observed a certain decrease in ionic conductivity under a compressive condition [11].

The mechanical effect on the conductivity must be analyzed from the viewpoint of practical applications such as to SOFCs and it could give great insight into developments of ionic conductivity materials. In this study, the mechanical viscoelasicity and the ionic conductivity of polycrystalline zirconia stabilised with 8 mol% yttrium were investigated. The ionic conductivity was also examined under the mechanical tensile-loading conditions. Based on the results, the mechanical effect on the conductivity was discussed in terms of the oxygen ion mobility.

EXPERIMENT

Zirconia polycrystalline samples stabilized with 8 mol% yttrium were prepared from a starting powder (8YSZ) (TZ-8Y, Toso). The pressed sample was sintered at 1873 K for 10 h. The surface of the sample was observed by a scanning electron microscope (SEM) (VE-8800, Keyence) after coated with gold in order to observe a grain size. A magnification of up to 2000 was used with an accelerating voltage of 2 kV. An impedance analysis with alternating current (AC) was conducted to examine ionic conductivity, σ, in 8YSZ. The sample was 1.7 mm in width and 0.63 mm in thickness, respectively. A platinum paste was used for the electrodes and the platinum wires wound around and connected to a LCR meter (3522-50, Hioki). The temperature ranged from 573 K to 1273 K and the frequency sweep ranged from 1 Hz to 100 kHz. The dynamic mechanical thermal analysis (DMTA) was conducted in order to examine the dynamic modulus, E, using a dynamic mechanical analyzer (Tritec 2000, Triton Technology). The mechanical loss, Q^{-1}, was also obtained using a single-cantilever mode. The geometry of the sample was the same as that used in the AC impedance analysis. The clamp span was 50 mm and the amplitude of 0.005 mm was applied to the sample. The test temperature ranged from 173 K to 673K and the frequency ranged from 0.316 Hz to 31.6 Hz. The AC impedance analysis at 663 K ($\pm$0.5 K) under static tensile-loading condition was carried out to investigate the effect of mechanical load on the ionic conductivity. The applied tensile load was under 2 N so as to give elastic deformation to the sample. The test was conducted at a constant temperature of 663 K.

DISCUSSION

AC Impedance analysis

The SEM observation demonstrated that the polycrystalline structure was homogeneous. The grain size was approximately 12.4 μm on average and the grain boundaries were clear.

Figure 1 shows an Arrhenius plot of the grain conductivity, σ, obtained from the AC impedance analysis. The σ monotonically increased with temperature, showing a slight curvature, and reached 0.146 S/cm at 1273 K. Activation energy, ΔH_c, for the ionic conductivity is given by:

$$\sigma T = A \exp (- \Delta H_c / k_B T) \qquad (1)$$

where T is temperature, A is the pre-exponential factor, and k_B is the Boltsmann's constant. The ΔH_c at higher temperature (from 923 K to 1273 K), which is known as the migration enthalpy [12], was 0.78 eV, and the one at lower temperature (from 523 K 923 K), which is known as a sum of the migration enthalpy and the association enthalpy [12], was 0.99 eV, respectively.

Dynamic mechanical thermal analysis

Figure 2 shows the DMTA results measured at a frequency of 1 Hz. The dynamic modulus, E, gradually decreased as the temperature increased. The E at 296 K (room temperature) was about 150 GPa and it decreased by half, approximately 75 GPa, at 673 K. This trend was the same as the previously reported for stabilized zirconia [5]. Since the modulus has been also reported to remain almost constant from 673 K to 1273 K, the modulus E in this study would be probably constant at around 75 GPa until 1273 K.

The results for the mechanical loss, Q^{-1}, revealed two peaks which appeared around 440 K and 540 K as shown in Fig. 2. This agreed well with the results obtained by other methods [6-9] while the present measurement was conducted by the DMTA. The activation energy for the mechanical relaxation process, ΔH_m, is given as follows [13]:

$$\Delta H_\mathrm{m} = k_\mathrm{B}\, \mathrm{d}(\log f)/\mathrm{d}(1/T_\mathrm{p}), \tag{2}$$

where T_p is a peak temperature of Q^{-1} measured at a frequency of f. The ΔH_m for the relaxation at lower temperature, 0.84 eV, was much smaller than that at higher temperature, 2.8 eV. These peaks are known to be attributed to different relaxation mechanisms [6-9]. The first peak of Q^{-1} at around 400 K could be due to simple complexes of an oxygen vacancy with dopant cations, i.e. $(Y_{Zr}'V_O^{\cdot\cdot})^{\cdot}$, as expressed in Kröger-Vink notation, e.g. for M_S^C, M, S, and C corresponds to the species, the lattice site that the species occupies, and the electric charge of the species relative to the site that it occupies, where $'$, $\cdot$, and $^\times$ indicate an electron, a vacancy, and a neutral charge, respectively. The other one at around 550 K could be assigned to either relaxation of oxygen vacancies within a couple (or a cluster) of dopant cations, i.e. $(2\,Y_{Zr}'V_O^{\cdot\cdot})^\times$, or that of dopant cations. In addition, ΔH_m for the relaxation at lower temperature was close to ΔH_c, so that the migrations of the complexes of oxygen vacancies with dopant cations would be dominant for the ionic conduction in 8YSZ. It should also be noted that, in a term of the viscoelasticity, E and Q^{-1} were linked each other as to relaxation phenomena, which implies that when the oxygen mobility is high, namely, the ionic conductivity is high, the modulus inevitably becomes low.

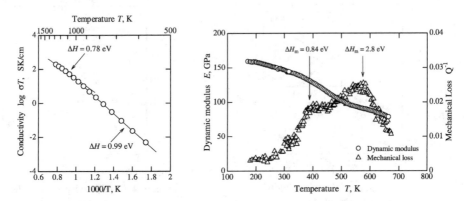

Figure 1 Arrhenius plots of conductivity

Figure 2 Dynamic mechanical properties at 1 Hz

AC impedance analysis under tensile stress

The pre-experimental static tensile test proved that the sample showed elastic deformations under the conditions that the temperature was from 296 K to 663 K and the load was below 2 N, i.e. the stress was below approximately 2 MPa. Thus, to examine the effect of the elastic load on the ionic conductivity, the following AC impedance test was conducted at 663 K with tensile load varying from 0 N to 2.0 N.

Figure 3 shows a Cole-Cole plots obtained from the AC impedance test under various tensile loadings. The ionic conductivity clearly depended on the load, i.e. the stress. The relationship between the ionic conductivity and the stress is shown in Fig. 4, where σ is normalized by the one without a load, σ_0. The normalized conductivity, σ/σ_0, monotonically increased with the load and enhanced by approximately 6% at maximum when the stress was approximately 2.0 MPa. It should be noted here that the effect of the change in the sample geometry on the conductivity was negligible because the elastic deformation was quite small as shown in Fig. 4. Additionally, it was observed that the improved conductivity was remained after the load was promptly removed.

The improvement in the ionic conductivity may be attributed to the enhanced migration of oxygen vacancies in the complexes, $(Y_{Zr}'V_O^{\cdot\cdot})^{\cdot}$, during the elastic deformation. As explained in the previous section, the migration of oxygen vacancies is a dominant factor for the ionic conductivity in 8YSZ and also the activated migration was mechanically detected, appearing as the mechanical loss in the DMTA results. In addition, this increase could be regarded as a temporary effect caused by the elastic deformation, because the improved conductivity was remained after the prompt unloading and it gradually changed probably by the relaxation phenomena after the complete unloading. Further experimental results and discussion will be reported in detail elsewhere. Therefore, the conductivity can be changed by the mechanical way such as stress and strain, while the change by the elastic deformation might be a temporary effect.

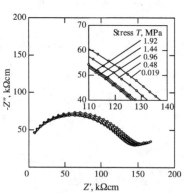

Figure 3 Cole-Cole plots
(measured at 663 K)

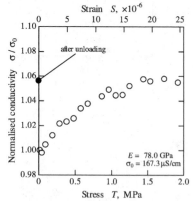

Figure 4 Conductivity and tensile stress
(measured at 663 K)

CONCLUSIONS

In this study, the mechanical effect on the oxygen mobility in polycrystalline zirconia stabilized with 8 mol% yttria (8YSZ) was investigated.

The ionic conductivity was evaluated by the impedance analysis with alternating current (AC). Also the dynamic modulus and mechanical loss were examined by the dynamic mechanical analysis (DMTA) and the activation energy for the mechanical relaxation was determined. The modulus gradually decreased with increasing temperature, especially around 400 and 550 K, where the mechanical loss had peaks at these temperatures. Both peaks had different activation energies due to different relaxation mechanisms. The comparison of the activation energies for the ionic conductivity to that for the mechanical relaxation proved that the oxygen mobility in 8YSZ was attributed to the immigration of an oxygen vacancy with a dopant cation in a form of simple complexes. The results also illustrated the strong relationship between the modulus and the oxygen mobility, namely the ionic conductivity.

The AC impedance analysis under mechanical tensile loads was carried out to examine the effect of the mechanical stress on the ionic conductivity. The tensile load improved the ionic conductivity by 6 % at maximum, and the improvement in the conductivity was remained after the load was promptly removed. The reason for the increase in the conductivity could be that the oxygen vacancies in the simple complexes were temporarily activated during the elastic deformation. The results of the present study proposed that the oxygen mobility in electrolyte of solid oxide fuel cell could be improved by mechanical means and also suggested that the coupling problem of the mechanical stress and ionic conduction must be considered, allowing for the relation between the mechanical properties and the ionic conductivity.

ACKNOWLEDGMENTS

We would like to acknowledge the Mizuho Foundation for the Promotion of Sciences and the Yazaki Memorial Foundation for Science and Technology for their financial supports.

REFERENCES

1. V.V. Kharton, F.M.B. Marques, A. Atkinson, *Solid State Ionics* **174**, 135-49 (2004).
2. S.P.S. Badwal, F.T. Ciacchi, *Ionics* **6**, 1-21 (2000).
3. F. Wakai, S.Sakaguchi, Y. Matsuno, *Adv. Ceram. Mater.* **1**, 259-263 (1986).
4. D.J. Green, R.H.J. Hannink, M.V.Swain, *Transformation Toughening of Ceramics* (CRC Press, 1989).
5. J.W. Adams, *J. Am. Ceram. Soc.* 80, 903-8 (1997).
6. J. Kondoh, H. Shiota, *J. Mater. Sci.* 38, 3689-94 (2003).
7. M. Ozawa, T. Itoh, E. Suda, *J. Alloys and Compounds* 374, 120-3 (2004).
8. M. Weller et al., 175, 329-33 (2004).
9. M. Weller et al, *Solid State Ionics* 175, 409-13 (2004).
10.Otsuka K et al, Appl. Phys. Lett. 82, 877 (2003).
11.JC M'Peko et al, *Solid State Ionics* 156, 59-69 (2003).
12.Y. Arachi et al, *Solid State Ionics* 121, 133-139 (1999).
13.A.S. Nowick and B.S. Berry, *Anelastic Relaxation in Crystalline Solids* (Academic Press, 1972).

Mater. Res. Soc. Symp. Proc. Vol. 1041 © 2008 Materials Research Society   1041-R03-10

Effect of Water Vapor and SOx in Air on the Cathodes of Solid Oxide Fuel Cells

Seon Hye Kim[1], Toshihiro Ohshima[1], Yusuke Shiratori[1], Kohei Itoh[1], and Kazunari Sasaki[1,2]

[1]Faculty of Engineering, Kyushu University, Fukuoka, 819-0395, Japan

[2]Hydrogen Technology Research Center, Kyushu University, Fukuoka, 819-0395, Japan

ABSTRACT

The influence of chemical species, water vapor and SOx, included in air on the voltage drop of SOFCs was studied. La_2O_3 was formed on the surface of cathode and caused the significant voltage drop of cells under high water vapor concentrations at 800°. The voltage drop of cells also occurred with SO_2 in air but was recovered with supplying SO_2-free air into the cells. It was considered that the cell voltage drop caused by SOx was due to the adsorption of sulfur and/or the formation of sulfur compounds on the surface of cathodes.

INTRODUCTION

Global warming is one of the serious environmental problems in these days and CO_2 gas produced from the oxidation of fossil resources is a main cause of the warming. Solid oxide fuel cells (SOFCs) are the promising energy system which can reduce the emission of CO_2 gas. SOFCs offer a clean, pollution-free technology to generate electricity with high efficiencies [1]. Therefore, there is a strong desire for the commercialization of SOFCs but there are many issues to be solved for commercialization. For the cathode of SOFCs, the ionization of oxygen on a cathode and the diffusion of oxygen ion in a cathode are very important, and the degradation mechanism of cathode during operation should be understood. It was reported that gaseous CrO_3 vaporized from interconnects made of chromium-containing alloys damaged the performance of cathode because of the precipitation of nonconductive Cr_2O_3 particles at triple phase boundaries [2]. It is believed that continuous exposure of cathodes to ambient air containing SO_2, NO_2, H_2S, H_2O and O_3 might causes cell performance degradation [3-4]. However, the influences of such minor constituents in air on the performance of cathodes have not been investigated yet intensively.

Therefore, the aim of the present work is to study the influence of water vapor and SOx contained in air on the performance of SOFCs and to find the degradation mechanism of cathodes. Power generation characteristics of the cells were analyzed with measuring cell voltage at a constant current density under supplying the artificially-contaminated air with water vapor or

SO$_2$ at various operation temperatures. The evolution of surface morphology and the change of chemical composition of the cathode surfaces due to poisoning were analyzed.

EXPERIMENTAL DETAILS

SOFCs with electrolyte-supported structure and anode-supported structure were used in this study. Composition of electrolyte with a diameter of 20 mm (Daiichi Kigenso Kagaku) was 10 mol% Sc$_2$O$_3$-1 mol%CeO$_2$-89% ZrO$_2$ (abbreviated by ScSZ). High purity 56% NiO-44% ScSZ powder mixtures were used for anodes. The cathode had a two-layer structure. Powder mixture of (La$_{0.8}$Sr$_{0.2}$)$_{0.98}$MnO$_3$ (LSM) and ScSZ with a weight ratio of 1 was applied to the first layer. LSM powder sintered at 1400° for 5 hrs was used for the second layer as a cathode current-collecting layer. The anode and cathode layers were prepared via screen printing followed by sintering at 1300° for 3 hrs and at 1200° for 5hrs, respectively. Electrode area was 8x8 mm^2. The cathode layer was also screen printed with the same method on the anode-supported cells.

For electrochemical characterization, H$_2$ fuel gas with water vapor (97% H$_2$-3%H$_2$O) was supplied to the anode. While the air with different water vapor concentrations or with different SO$_2$ concentrations was supplied to the cathode, cell voltages at the constant current of 200 mA/cm^2 were measured at various operation temperatures. High-resolution FESEM (S-5200, Hitachi), HRTEM (JEM-2000EX/T) were used to analyze microstructural changes caused by air impurity. Mg Kα radiation from a monochromatized X-ray source was used for XPS measurements (ESCA-3400, KRATOS).

RESULTS AND DISCUSSION

Influence of water vapor on the cathode of SOFCs

Figure 1 shows the cell voltage drop as a function of water vapor concentration at (a) 800° and (b) 1000°. It increased with increasing water vapor concentration especially at 800°. It rose to 270 mV at 800° but to 45 mV at 1000°. It should be noticed that there was no voltage drop in the cell tested with 3 vol% wet air both at 800 and 1000°.

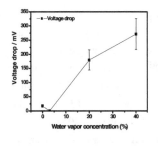 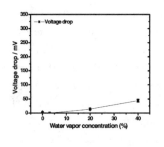

(a)   (b)

Figure 1. Dependence of cell voltage drop on water vapor concentration (a) at 800° during 2.5 hrs and (b) at 1000° during 5 hrs, at 200 mA/cm^2 for the cells with ScSZ electrolytes.

The cell performance under the 3 vol% wet air was even better than that under the dry air. This result is consistent with the study by Sakai et al.[5] who reported that the interfacial conductivity increased by adding a small amount of water vapor to the air. From these results, it can be concluded that high humidity makes a serious degradation of cathode, especially at 800°.

Figure 2 shows the changes of surface morphology of cathodes tested with wet air at 800°. The cathode consisted of two layers as described above; the first layer was made up of LSM and ScSZ and the second layer LSM only. There were no changes in the first layer before and after the test at 800 and 1000°. The surface of the second layer also showed negligible change for cells tested under dry air as shown in Figure 2(a). However, a lot of small particles were formed on the surface of the second layer tested under wet air as shown Figure 2(b). In the case of 40 vol% wet air, LSM surface was covered with small particles, as shown in Figure 2(c). The changes of LSM surface morphology were associated with the cell voltage drop.

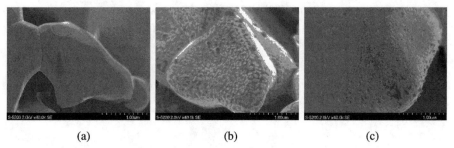

(a)   (b)   (c)

Figure 2. Micrographs of $(La_{0.8}Sr_{0.2})_{0.98}MnO_3$ cathodes, after the electrochemical test for 5 hrs (200 mA/cm^2, at 800 ℃) under the flow of (a) dry air, (b) with 20 vol% H$_2$O, and (c) with 40 vol% H$_2$O.

In the XP spectra, La, Sr, Mn and O elements were detected on the cathode surfaces. The O1s core level XP spectra of cathode materials are shown in Figure 3. After annealing in dry air, we found only one emission peak at 530.9 eV, as shown in Figure 3(a). In the case of wet air, however, we found 2 peaks overlapped. After annealing in 20 vol% wet air, the emission peak 532.7 eV became predominant and the peak at 530.3 eV could be distinguished as shown in Figure 3(b). The former peak can be assigned to O^{2-} ion of La_2O_3 located at 532.8 eV and the latter peak to O^{2-} ion of MnO (529.9 eV) and SrO (530.2 eV). In the case of 40 vol% wet air, we could again find 2 peaks located at 533.2 eV and 531.1 eV. Our XPS measurement using pure La_2O_3 also revealed similar 2 signals: a larger peak at 533.8 eV and a smaller peak at 531.2 eV, indicating that the cathode surface may be covered with La_2O_3[6]. We could also find La_2O_3 on the LSM with 40% wet air at 800°C in TEM analysis[7].

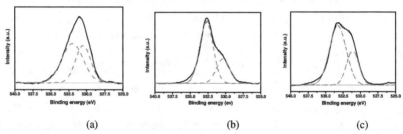

(a)   (b)   (c)

Figure 3. The O1s core level XP spectra of $(La_{0.8}Sr_{0.2})_{0.98}MnO_3$ annealed in (a) dry air, (b) air -20   vol% H_2O, and (c) air - 40 vol% H_2O.

We could therefore conclude that the particles segregated on the surface of the second layer were La_2O_3 from the results of XPS and TEM. La_2O_3 has poor electrical conductivity. Therefore, the formation of such insulating particles on the cathode surface may lead to irreversible cell voltage drop.

Poisoning of SOx on the cathode

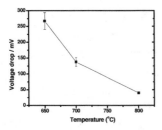

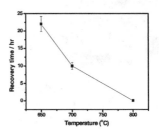

Figure 4. Temperature dependence of cell voltage drop at 200 mA/cm^2 for 5 hrs using SO$_2$ (20 ppm) in air.

Figure 5. Recovery time without SO$_2$ after poisoning test for 5 hrs at 200 mA/cm^2 (SO$_2$ conc.= 20 ppm).

Figure 4 shows the change of cell voltage drop as a function of temperature in cells tested for 5hrs under a SO$_2$ concentration of 20 ppm at the current density of 200 mA/cm^2. Figure 5 shows time required to fully recover the voltage drop shown in Figure 4 as a function of temperature after cutting off SO$_2$. The voltage drop linearly increased with decreasing operation temperature. From these results, it is clear that higher SO$_2$ concentration causes more severe poisoning of cathode especially at temperature below 800°.

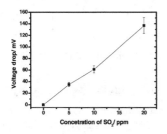

Figure 6. SO$_2$ concentration dependence of cell voltage drop at 700° and 200 mA/cm^2 during 20 hrs in anode-supported cells.

Figure 6 shows the change of cell voltage drop as a function of SO$_2$ concentration in anode-supported cells tested for 20 hrs at 700° and the current density of 200 mA/cm^2. Under this condition, the voltage drop linearly increased with increasing SO$_2$ concentration by approximately 7 mV/ppm. These results mean that the influence of SO$_2$ becomes significant with decreasing operation temperature.

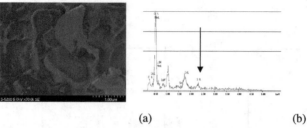

(a)   (b)

Figure 7. (a) FESEM micrograph and (b) EDX results of $(La_{0.8}Sr_{0.2})_{0.98}MnO_3$ cathodes, after SO_2 poisoning for 20 minutes (200 mA/cm^2) at 700°.

Figure 7(a) shows the surface morphology of the cathode tested at 700° for 20 min and then rapidly cooled under the flow of air with SO_2, and Figure 7(b) shows the identified elements in the area of Figure 7(a) using EDX. There was little difference in surface morphology before and after cell testing, but we could find sulfur traces on the surface of the cathode after testing. We could also find S 2p peak on the surface of the LSM with 20 ppm SO_2 in air at 700° in XPS analysis. The chemical adsorption of sulfur and/or the formation of sulfur compounds on the cathode surface may lead to a cell voltage drop and to poison the cathode reactions because they reduce the active electrode reaction area (triple phase boundaries) for the reduction of oxygen. A longer term experiments will be needed to understand the cathode degradation phenomena caused by such poisoning.

CONCLUSIONS

The influence of water vapor and SOx in air on the performance of SOFCs was examined and the following conclusions could be obtained:
1. Reversible cell voltage change was observed at 1000° while irreversible degradation occurred at 800° with very high water vapor concentrations. It was revealed that the degradation was mainly due to the formation of La_2O_3 segregated on the surface of $(La_{0.8}Sr_{0.2})_{0.98}MnO_3$ cathodes.
2. SO_2 contained in air caused a voltage drop of SOFCs cells, but the voltage drop could be recovered with supplying a SO_2-free air into the poisoned cells. A large cell voltage drop was observed at operation temperatures below 800°, while a little voltage drop was measured at above 800°. It was believed that the cell voltage drop was due to the adsorption of sulfur and/or the formation of sulfur compounds on the cathode surfaces.

ACKNOWLEDGMENTS

This study was partly supported by the NEDO SOFC R&D project. The authors also gratefully acknowledge the 21th Century of Excellence (COE) program.

REFERENCES

1. S.C. Singhal, *Solid State Ionics*, **135**, 305-313 (2000).
2. S. Taniguchi, M. Kadowaki, H. Kawamura, T. Yasuo, Y. Akiyama, Y. Miyzke, T. Saitoh, *J. Power Sources,* **55** 73 (1995).
3. R. Mohtadi, W.-K. Lee, and J.W. Van Zee, *J. Power Sources,* **138** 216-225 (2004).
4. K. Sasaki, S. Adachi, K. Haga, M. Uchigawa, J. Yamamoto, A. Ioshi, J.-T. Chou, Y. Shoratori, and K. Itoh, *ECS Transac*, **7** (1) 1675-1683 (2007).
5. N. Sakai, K. Yamaji, T. Horita, Y.P. Xiong, H. Kishimoto, H. Yokokawa, *J. Electrochem.Soc*, **150** (6), A689 (2003).
6. B. V. Crist, "Handbooks of Monochromatic XPS Spectra" vol.5, (1999).
7. S.H. Kim, K.B. Shim, C.S. Kim, J.-T. Chou, T. Oshima, Y. Shiratori, K. Itoh, and K. Sasaki, *J. Fuel cell Sci. Technol.*, submitted.

Mater. Res. Soc. Symp. Proc. Vol. 1041 © 2008 Materials Research Society   1041-R03-11

Compaction and Cold Crucible Induction Melting of Fine Poly Silicon Powders for Economical Production of Polycrystalline Silicon Ingot

Daesuk Kim[1], Jesik Shin[2], Byungmoon Moon[2], and Kiyoung Kim[1]

[1]Material Engineering, Korea University of Technology and Education, Byung-cheon, Cheon-an, 330-708, Korea, Republic of

[2]Korea Institute of Industrial Technology, Songdodong7-47, In-cheon, 406-800, Korea, Republic of

ABSTRACT

The consolidation and casting processes of fine silicon powders, by-product of high purity silicon rods making process in the current method, were systematically investigated for use as economical solar-grade feedstock. Morphology, size, and contamination type of the fine silicon powders were inspected by combined analysis of SEM, particle size analyzer, and FT-IR. Silicon powder compacts were tried to fabricate by a consolidation process without a binding agent and then their density ratio and strength were evaluated. Finally, the electrical resistivity of the specimens prepared by an electromagnetic casting method was examined for purity assessment.

INTRODUCTION

The annual world photovoltaic (PV) market entered the age of GW, in a rapid expansion with the average annual growth rate of 35% since the middle 1990s. The strongly growing PV market is based on crystalline silicon technology. More than 90% of the annual solar cell production is made of crystalline silicon wafers. The issues which the PV industry is currently facing are the cost reduction and the shortage of solar grade silicon due to strongly increasing demand; Around 70% of the production cost of solar module comes from silicon wafers, and the shortage of silicon supply is limiting the growth of the worldwide PV-industry based on silicon [1,2].

Most of the existing facilities for high purity silicon production are using the established Siemens process, as shown in figure 1. In the conventional Siemens process, purified trichlorosilane (TCS; SiHCl$_3$) gas is reduced by hydrogen and deposited on silicon or tantalum rods. But, this technology provides highly pure but very expensive silicon due to low deposition yield into silicon rods. Therefore, in this study, the consolidation and casting processes of the low-priced silicon powders, by-product of high purity silicon making process in the current method, were systematically investigated to economically produce a multi-crystalline silicon wafer for solar cell. The characteristics of the silicon powders were inspected by combined analysis of SEM, particle size analyzer, and FT-IR. Silicon powder compacts were tried to fabricate by a consolidation process without binding agents. Also, the melting experiment and electrical resistivity examination were carried out for purity assessment.

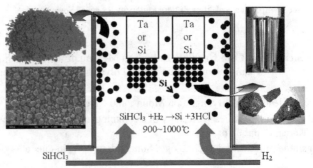

Figure 1. Schematic of a conventional Siemens process for high purity silicon production.

EXPERIMENTAL

Morphology, size, and contamination type of the fine silicon powders were inspected by combined analysis of SEM, particle size analyzer, and FT-IR. In order to form the silicon powders into a disc with 15 mm diameter, firstly, they were etched in a concentrated HF solution, a solution of 40% aqueous HF:ethanol (50:50), for removing surface silicon dioxide. After subsequent rinsing with deionised water and ethanol to remove trace HF, the solutions were filtered through a Buchner vessel (with a 0.1 μm filter paper). Then the powders were dried in a vacuum oven of 60°C for 10 hours. The dried powders were immediately loaded into an ASTM A681 die in a vacuum chamber. The dried poly silicon powders were subjected to uniaxial mechanical pressing at pressure in the range of 70~300 MPa for 5 min. All pressing was carried out at room temperature under a vacuum of 10^{-3} torr. Simultaneously, untreated powders were also compacted to compare directly the effects of the HF pre-conditioning. For the evaluation of compact strength, the fabricated discs were shaken in No.16 sieve for 30 min, and then weight loss was measured. The amplitude of vibration was 0.14~0.55 mm and frequency was 60 Hz. Also, the discs were dropped on an ASTM A270 plate, and the number of broken pieces was counted. Finally, to examine the applicability as solar grade feedstock material, the silicon powder compacts were melted by electromagnetic casting (EMC) method using a cold crucible [3], and the electrical resistivity was examined for purity assessment utilizing Hall effect measurement.

RESULTS AND DISCUSSION

Characteristics of silicon powder

The characteristics of the silicon powders, produced as by-product in Siemens process, were inspected using SEM, particle size analyzer, and FT-IR spectrometer. Those were close to spherical shape and their average diameter was 7.8 μm and the mode 10.3 μm, as shown in figure 2. FT-IR spectrum result reveled that the main contaminants of the fine silicon powders were SiO_2 type oxide and humidity. It seems that the fine powders are readily oxidized and contaminated with moisture due to their large surface area.

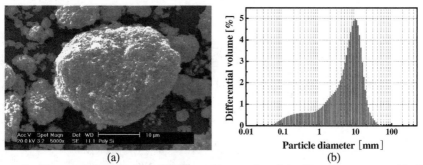

(a)	(b)

Figure 2. Characteristics of the silicon powders produced in Siemens process; (a) SEM morphology and (b) particle size distribution.

Compactablity of silicon powder

For the use as feedstock material in PV industry, the silicon powders have to be modified into a suitable shape for easy handling, transport, and charging as well as purified up to solar grade. Therefore, the applicability of a binder-free consolidation process of the silicon powders was systematically investigated. For consolidation without binding agents, firstly, the silicon powders were etched in the concentrated HF-ethanol solution, which was reported to be effective for removing oxide on silicon surface [4]. After subsequent rinsing with de-ionised water and ethanol, the solutions are filtered through a Buckner vessel and then the powders were dried in a vacuum atmosphere. Finally, the dried poly silicon powders were formed into disc by uniaxially pressing up to 300 MPa at room temperature. Figure 3 shows the representative silicon powder compacts fabricated in this study.

Figure 3. Photographs of silicon powder compacts (a) without and (b) with cracks.

Cracks were frequently observed in the process of removing the compacts from the die. Figure 4 shows the yield of consolidation process of the silicon powders as functions of processing atmosphere and applied pressure. It is interesting to note that the yield remarkably increased by 20% under a vacuum condition compared to that of ambient condition. This phenomenon seems to be related to how effectively silane gas, explosive in air, was able to be removed. Barraclough et al. reported that silane was detected when untreated silicon powders were milled in air and when HF pre-conditioned powders were compacted and exposed to air [5,6]. And above 200 MPa, the yield was deteriorated due to non-uniform distribution of residual stress in the compacts. It was confirmed by crack analysis using a commercial software, MSC/MARC, and a cone-like fracture pattern observed in the actual experiment.

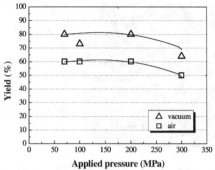

Figure 4. Yield of silicon powder compacts as a function of applied pressure and consolidation atmosphere.

The applicability of the silicon powder compacts as feedstock material for PV crystalline silicon ingot production was evaluated by the falling and shaking tests to simulate the actual handling and charging conditions in casting process. The comparably higher strength was achieved by the chemical pre-treatment using the concentrated HF solution prior to consolidation process, as shown in figure 5. Also, the density ratio, which is normalized with theoretical bulk density of silicon (2.33 g/cm^3), reached approximately 70% in crack-free consolidation condition (200 MPa) after the chemical pre-treatment. It is likely that hydrogen-terminated surface of silicon powders caused by the chemical pre-treatment promoted bonding between silicon powders. This is in accord with other's results; the electrical continuity and packing density of silicon powder compacts were improved by chemical pre-conditioning using a HF solution [5].

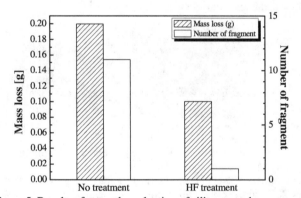

Figure 5. Results of strength evaluation of silicon powder compacts.

Melting of silicon powder compacts

In order to examine whether the silicon powder compacts is able to be used as solar grade feedstock material or not, they were melted on semiconductor grade FZ silicon substrate under non-contact condition with crucible wall utilizing EMC method [3], and then electrical resistivity was measured. The specimen preparation condition used in this melting experiment was summarized in table 1. Wet chemical treatment designates pre-conditioning etching, rinsing, and drying processes using the HF solution, which is explained in experimental part.

Table 1. Specimen preparation conditions for melting experiment using silicon powders

No.	Drying before cleaning	Wet chemical treatment	Compaction	Dry chemical treatment
1	x	x	x	x
2	x	O	x	x
3	x	O	O	x
4	O	x	x	x
5	O	x	x	O

Figure 6 shows the variation of electrical resistivity with various silicon powder processing conditions. The electrical resistivity of specimen 1, which was prepared by melting the as-received silicon powders, was 0.22 Ωcm. It means that the silicon powders, by-product of the current Siemens process, was not able to be used alone as solar grade feedstock material, because the required value for a solar cell is typically above 0.5 Ωcm. In specimen 3, which was prepared by melting the silicon powder compacts, electrical resistivity decreased below 0.1 Ωcm. It seems to be because reaction products generated in the wet chemical treatment and the used chemical solutions were not completely removed from the compacts. On the other hand, when the silicon powders were dried (No 4 on table 1) or reductively heat-treated at 1,200~1,300°C for 1 hour under a 10%H_2-Ar atmosphere (No 5 on table 1), electrical resistivity increased up to 3~4.6 Ωcm.

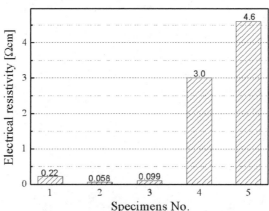

Figure 6. Variation of electrical resistivity as a function of powder processing condition.

CONCLUSION

The consolidation and non-contact electromagnetic melting of the low-priced silicon powders, by product of the current Siemens process, were systematically investigated to develop the production technology for a cheap solar grade feedstock material. The silicon powders were close to spherical shape and their average diameter was 7.8 μm. The main contaminants of the fine silicon powders were SiO_2 type oxide and humidity. The yield of the binder-free consolidation process increased by 20% under a vacuum condition compared to ambient condition. The chemical pre-treatment using the HF solution was observed to be effective for the improvement of the density ration and strength of the silicon powder compacts. The silicon powders in as-received state were not pure enough to be used alone as solar grade feedstock material. After the adequate chemical treatments, a sufficiently high purity above solar-grade was able to be achieved.

REFERENCES

1. M. Schmela, *Photon International* **March**, 66 (2005).
2. A. Müller, M. Ghosh and R. Sonnenschein, P. Woditsch, *Mat. Sci. & Eng* **B 134**, 257 (2006).
3. J. S. Shin, H. S. Kim, S. M. Lee and B. M. Moon, *Materials Science Forum* **475-479**, 2671 (2005).
4. V. Palermo and D. Jones, *Materials Science in Semiconductor Processing* **4**, 437 (2001).
5. K. G. Barraclough, A. Loni, E. Caffull and L. T. Canham, *Materials Letter* **61**, 485 (2007).
6. I. Lampert, H. Fuzstetter and H. Jacob, *J. Electrochem. Soc.* **133**, 1472 (1986).

Energy LCA Methodology

Mater. Res. Soc. Symp. Proc. Vol. 1041 © 2008 Materials Research Society   1041-R05-01

Life Cycle Assessment of Future Fossil Technologies with and without Carbon Capture and Storage

Roberto Dones[1], Christian Bauer[1], Thomas Heck[1], Oliver Mayer-Spohn[2], and Markus Blesl[2]
[1]Paul Scherrer Institut, Villigen PSI, 5232, Switzerland
[2]IER Universität Stuttgart, Stuttgart, 70565, Germany

ABSTRACT

The NEEDS project of the European Commission (2004-2008) continues the ExternE series, aiming at improving and integrating external cost assessment, LCA, and energy-economy modeling, using multi-criteria decision analysis for technology roadmap up to year 2050. The LCA covers power systems suitable for Europe. The paper presents environmental inventories and cumulative results for selected representative evolutionary hard coal and lignite power technologies, namely the Ultra-Supercritical Pulverized Combustion (USC-PC) and Integrated Gasification Combined Cycle (IGCC) technologies. The power units are modeled with and without Carbon Capture and Storage (CCS). The three main technology paths for CO_2 capture are represented, namely pre-combustion, post-combustion, and oxy-fuel combustion. Pipeline transport and storage in geological formations like saline aquifers and depleted gas reservoirs, which are the most likely solutions to be implemented in Europe, are modeled for assumed average conditions. The entire energy chains from fuel extraction through, when applicable, the ultimate sequestration of CO_2, are assessed, using ecoinvent as background LCA database.

The results show that adding CCS to fossil power plants, although resulting in a large net decrease of the CO_2 effluents to the atmosphere per unit of electricity, is likely to produce substantially more GHG than claimed by near-zero emission power plant promoters when the entire energy chain is accounted for, especially for post-combustion capture technologies and hard coal as a fuel. Besides, the lower net power plant efficiencies lead to higher consumption rate of non-renewable fossil fuel. Furthermore, consideration of the full spectrum of environmental burdens besides greenhouse gas (GHG) results in a less definite picture of the energy chain with CCS than obtained by just focusing on GHG reduction.

INTRODUCTION

Fossil power remains crucial for covering a substantial part of the steadily increasing power demand in advanced countries as well as the dramatically increasing energy demand of fast developing countries. Therefore, improvements in fossil power technology likely to be implemented on a large scale in the next decades are important for contributing to control the emissions of greenhouse gas (GHG) by substitution of obsolete plants by highly efficient units, refurbishment of aging plants, and CO_2 sequestration. An equal installation rate from renewable sources or nuclear seems not realistic in the short to medium term on a worldwide scale.

The paper presents preliminary Life Cycle Assessment (LCA) results for coal (being currently more complete than natural gas) obtained by the large European project NEEDS (New Energy Externalities Developments for Sustainability) (2004-2008). NEEDS is an integrated project financed under 6th RTD (Research, Technology Development and Demonstration) Framework Programme of the European Union (EU), and includes 66 partners (Universities, Research Institutions, Industry, NGOs), representing 26 Countries within and outside the EU [1].

The main objective of the NEEDS project is to evaluate the long term sustainability of energy technology options and policies for Europe in the first half of the 21st century, both at the level of individual countries and for the enlarged EU as a whole. This will be achieved by means of full costs and benefits (i.e. direct and external) assessment of future energy systems (in particular electricity) and energy policies.

The NEEDS project continues the ExternE series by the European Commission. The original aim is at improving and integrating three methodologies, namely external cost assessment, LCA, and energy-economy optimization modeling for electricity technology roadmap up to year 2050. Besides total cost assessment, Multi-Criteria Decision Analysis (MCDA) is applied in order to encompass aspects which cannot be merely monetized. Both methods will be used in parallel for holistic technology assessment.

METHODOLOGY

The main objective of the NEEDS research stream on LCA is to provide data on life cycle inventories and costs for a wide range of key emerging energy technologies, with focus on long term (2035-2050) technical developments. A link between LCA tools and energy system modeling (Markal-Times) has been established, enabling adequate feedback loops between future energy system configurations and the life cycle inventories (LCI) of individual energy technologies [1]. However, the preliminary results illustrated in this paper for coal systems only are obtained using a simplified calculation without application of the full spectrum of new features being developed within the project, which consists in the use of electricity mixes calculated by energy-economy model depending of scenarios and use of assumed future material processing for key (widespread) materials. However, these aspects are not much important for fossil systems, due to their characteristically high environmental burdens during operation of the power plant and the associated fuel supply chain.

The LCA research stream addresses power systems only. The level of detail in the LCA description of technologies is limited by the speculative nature of the assessment of medium to long term technologies. An evolutionary approach is applied on the basis of the LCI for current best available technologies. On the other hand, such approximate models are likely sufficient to serve the integration of different methodologies and estimate environmental effects (measured by external costs) of specific energy policies. Nevertheless, the results obtained should always be considered with care and the description of future technologies requires continuous refining.

Table I shows the net efficiencies assumed for natural gas and coal power plants with and without (w/o) CCS for the range of scenarios in year 2025 and 2050 defined in this study [3], after different literature for the range of efficiencies of power plant w/o CCS and after [2] for CO_2 capture (and compression) efficiency penalties. According to the definition agreed upon within the LCA research stream, the scenarios should reflect pessimistic (Pe), realistic-optimistic (RO), and very optimistic (VO) conditions anticipated for each of the technologies individually modeled. Therefore, assumptions made on different technologies (e.g., coal PC, photovoltaic, or nuclear generation IV) may not be consistent with each other in terms of energy policy scenario or total installed capacity. Nevertheless, the information gained by studying the range of likely values is very valuable to understand orders of magnitudes and identify key parameters to vary in future sensitivity analyses.

Table I. Net efficiencies assumed for natural gas and coal power plants with and without CCS for the range of scenarios in year 2025 and 2050 defined in this study [3].

Fuel	Conversion technology	Capture technology class (combustion)	Year	Sc. *)	Net electric efficiency w/o CCS %	Net electric efficiency w/ CCS %	Efficiency penalty **) %	CO_2 capture efficiency %
Natural gas	NGCC	Post	2025	Pe	61	53	8	90
				RO	62	56	6	90
				VO	63	57	6	90
			2050	Pe	62	56	6	90
				RO	65	61	4	90
				VO	66	62	4	90
		Oxyfuel	2025	Pe	61	51	10	100
				RO	62	52	10	100
				VO	63	53	10	100
			2050	Pe	62	52	10	100
				RO	65	60	5	100
				VO	66	61	5	100
Coal	PC	Post	2025	Pe	47	37	10	90
				RO	49	42	7	90
				VO	52	45	7	90
			2050	Pe	50	43	7	90
				RO	54	49	5	90
				VO	57	52	5	90
		Oxyfuel	2025	Pe	47	37	10	99.5
				RO	49	41	8	99.5
				VO	52	44	8	99.5
			2050	Pe	50	42	8	100
				RO	54	47	7	100
				VO	57	50	7	100
	IGCC	Pre (hard coal)	2025	Pe	53	47	6	90
				RO	54	48	6	90
				VO	55	49	6	90
			2050	Pe	53.5	47.5	6	90
				RO	54.5	48.5	6	90
				VO	55.5	49.5	6	90
		Pre (lignite)	2025	Pe	51	45	6	90
				RO	52	46	6	90
				VO	53	47	6	90
			2050	Pe	51.5	45.5	6	90
				RO	52.5	46.5	6	90
				VO	53.5	47.5	6	90

*) Sc.=Technology scenario; Pe=Pessimistic; RO=Realistic-optimistic; VO=Very optimistic.

    For the sake of internal consistency, the results reported herewith for all coal power plants (PC and IGCC) refer to a coal with the same characteristics of lower heating value (LHV), namely 26 MJ/kg for hard coal and 8.8 MJ/kg for lignite, and CO_2 emission per MJ fuel (92.2 g/MJ for hard coal and 108.3 g/MJ for lignite).

The air emissions from a modern, supercritical hard coal PC of 509 MW net capacity, and 43% net efficiency (Rostock, Germany) have been considered as reference also for future Ultra-Supercritical Pulverized Combustion (USC-PC) of 600 MW (w/o CCS). Reference for the current state-of-the-art lignite PC technology is the power plant "Niederaussem K" (Bergheim, Germany). The plant is named "BoA-unit" (Braunkohlekraftwerk mit optimierter Anlagentechnik), i.e. lignite plant with optimized systems engineering. The plant has a net capacity of 950 MWe and a net efficiency of 43.2%. The reference IGCC power plant considered in NEEDS for current conditions is an 'enhanced Puertollano (Spain) IGCC power plant', burning only hard coal, modified to higher net power (450 MWe instead of 300 MWe). The efficiency has been assumed 45% instead of the current effective 42.2% in Puertollano. LCI data for future USC-PC and IGCC units have been approximated using the above reference cases, as suitable approximation for an LCA study. Nevertheless, sensitivity analyses should be performed on the various key parameters.

Although the LCA for this study is for power plants to be built up to 2050, a simplified assumption consists in having no changes in mining, processing, and transport technology with respect to today's conditions as modeled in ecoinvent [4]. Although selected background materials (e.g. steel) and appropriate electricity mixes (depending on scenarios calculated by Markal-Times) will be modeled for the final integration in NEEDS, the expected influence on the results is minor except for some effects on the environmental characterization of electricity expenditures for pumping CO_2 in the deepest geological repositories.

For what concerns the chemicals used for CO_2 separation from the exhaust in post-combustion processes, using an amine-based solvent, the upper range of the values reported in [5] for amine (modeled with monoethanolamine, from the ecoinvent database [4]), NaOH, and active carbon (modeled with charcoal) have been assumed in the study. Additionally, the difference between the total chemical requirements for separation reported in [6] (2.76 kg/kWh) and the total materials from [5] has been modeled with generic chemicals organic, again from the ecoinvent database [4]. These values have not been changed for the different scenarios, because no information was available on process modifications in the future. This aspect would deserve a follow-up LCA modeling activity. Furthermore, in this preliminary study material needs for oxyfuel combustion could not be modeled, but only the energy uses for O_2 separation. Concerning emissions like NOx and SOx from coal PC, reductions along [6] have been considered, but adjusting SO_2 emission to 0.1 g/kWh thus assuming scrubber efficiency of about 99%, and using bound N content in the hard coal for oxyfuel combustion exhaust (giving <0.2 g/kWh).

The electricity used for CO_2 compression at the power plant is treated as auxiliary uses and consequently subtracted from the gross electricity generated. Therefore, the LCA boundary for the operation of power plants with CCS includes the separation and compression of the CO_2.

The modeling of the CO_2 transport and storage in supercritical state has been conducted on the basis of an engineering bottom-up modeling approach [7,8]. The transport is assumed to occur by pipeline with mass flow of 250 kg/s, which would correspond to roughly three hard coal power plants with CSS of the 500 MW class, as modeled in NEEDS. Two distances have been considered, 200 km and 400 km, the first without intermediate recompression, the second with one recompression (approximately 30 bar) after the first 200 km. Two cases have been modeled for deep geological storage: a saline aquifer at 800 m depth and a depleted gas field at 2500 m depth. The overpressure assumed additional to the hydrostatic pressure of the reservoir is about 30 bar for both cases (which may be high for aquifers). For each of the modeled reservoirs, two injection wells have been assumed, each with 125 kg CO_2 per second. These depths, the overpressure, and the injection rate per well have been defined on the basis of a survey of existing storage test projects [7]. They should represent reasonable values for applicable cases in Europe, but no assessment of the distribution of potential storage sites vs. (potential) sites for fossil power plants has been attempted in NEEDS. A key parameter for LCA is the electricity expenditures for compression of supercritical CO_2, calculated based on the realistic formula as function of the supercritical CO_2 mass flow from [9].

DISCUSSION

The results shown herewith include three aggregated indicators:
1. Greenhouse gas emission estimated using greenhouse warming potential by IPCC (2001), measured in kg(CO_2-equivalent)/kWh at busbar; the reason of this choice lies in the need to have figures on the expected reduction of total GHG from fossil systems with CCS;
2. Cumulative Energy Demand (CED) non-renewable using average CED for hard coal, lignite, natural gas, and uranium from ecoinvent [4], measured in MJ/kWh at busbar;
3. The life-cycle impact assessment (LCIA) method Eco-indicator'99 with hierarchist (H) perspective and average (A) weighting factors [10] EI'99(H,A), measured in eco-points; the reasons of the choice of this particular LCIA method lie in its widespread use within the LCA community as a measure of potential, overall environmental damages.

In particular, EI'99 [10] describes environmental effects for emissions occurring in Europe. It is a damage-oriented method, which considers, by means of damage factors, the effects of all emitted or used substances in three damage categories: Human Health, Ecosystem Quality, and Resources (fossil and mineral). The different damage categories are normalised, then further weighted on the basis of the perspective of three typologies of stakeholders, identified using a cultural theory concept: individualist, egalitarian, and hierarchist. The hierarchist has a balanced time perspective and requires consensus among scientists for inclusion of a burden; this perspective is considered the closest to the scientists' point of view.

External costs have not been calculated because the factual integration of LCA and external costs methodologies has not been performed so far.

Hard Coal and Lignite PC

Figure 1 shows synoptically the results per kWh at busbar obtained for the three aggregated indicators for the energy chains with hard coal (left hand side graphs) and lignite PC (right). The efficiency range assumed for the (reference) power plants w/o CO_2 capture for Pe-VO scenarios are well reflected by the fork of the corresponding results for all indicators.

Table II reports the GHG contribution analysis. GHG decreases for energy chains associated to PC with CCS from the equivalent chains for PC w/o CCS for different years (around 2010, 2025, and 2050) and scenarios range 63% to 84% for hard coal, and 75% to 94% for lignite systems. The lower values pertain to GHG decreases with oxyfuel combustion technology, because of higher CO_2 capture rate (up to nearly 100%) than for post-combustion and no consumption of chemicals as used for CO_2 separation from the exhaust in post-combustion CO_2 capture processes (as said, in this preliminary study the material needs for oxyfuel combustion could not be modeled, but only the energy uses for O_2 separation).

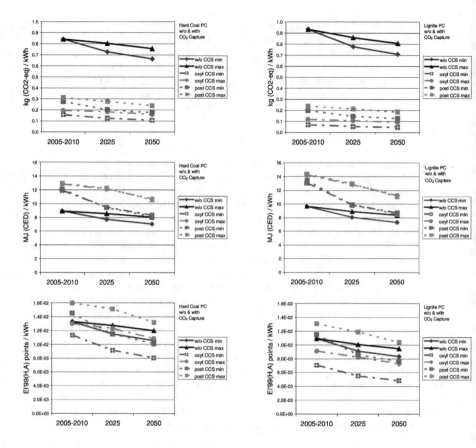

Figure 1. Greenhouse gas emission, Cumulative Energy Demand non-renewable, and EI'99(H,A) points (from top to bottom, respectively) results for PC hard coal and PC lignite (left and right, respectively) without CO_2 capture and with post- and oxyfuel-combustion CO_2 capture on each graph) and two cases for assumed European average CO_2 transport and injection depth (min = 200 km and 800 m depth aquifer; max = 400 km and 2500 m depth depleted gas field).

Table II. Cumulative and process-related GHG emission ranges for coal systems analyzed in this study [3].

GHG emission g(CO_2eq)/ kWh	fuel extraction & processing		power plant construction & dismantling		power plant operation (incl. CO_2 capture & compression)		CO_2 transport & storage		TOTAL		Year
	min	max	min	max	min	max	min	max	min	max	
Lignite, PC	16	16	2	2	916	916	0	0	934	934	2010
	13	15	2	2	762	843	0	0	777	859	2025
	12	14	2	2	695	792	0	0	708	808	2050
Lignite, PC, CCS	21	21	3	3	23	150	20	71	69	237	2010
	15	19	3	3	18	135	15	63	53	214	2025
	13	16	2	3	16	117	13	56	46	185	2050
Hard Coal, PC	84	87	2	2	737	753	0	0	823	842	2010
	75	83	2	2	651	720	0	0	727	805	2025
	68	78	2	2	594	677	0	0	664	757	2050
Hard Coal, PC, CCS	111	111	4	4	24	146	16	57	156	311	2010
	86	105	3	4	19	120	13	54	125	277	2025
	75	93	3	3	13	104	11	48	106	240	2050
Lignite, IGCC	16	16	1	1	903	903	0	0	920	920	2010
	13	13	1	1	750	779	0	0	764	793	2025
	13	13	1	1	743	772	0	0	757	786	2050
Lignite, IGCC,CCS	na		na		na		na		na		2010
	15	15	1	2	99	104	14	47	129	168	2025
	14	15	1	1	98	103	14	47	128	166	2050
Hard Coal, IGCC	87	87	6	6	749	749	0	0	842	842	2010
	71	74	1	1	613	636	0	0	685	711	2025
	70	73	1	1	608	630	0	0	679	704	2050
Hard Coal, IGCC, CCS	na		na		na		na		na		2010
	80	83	1	1	79	83	11	38	172	206	2025
	79	82	1	1	79	82	11	38	170	203	2050

Although lignite has a higher CO_2 formation per MJ fuel, the cumulative GHG emissions from power plants with CCS, with plant efficiency equal to hard coal PC by the same year and technology scenario, are lower because of the relatively low contribution of the upstream chain compared to hard coal.

The decreased net efficiency of the fossil power plant with CO_2 capture with respect to the same unit class and technology level w/o CO_2 capture leads to an increase of emissions of GHG in the fuel supply chain, proportional to the difference of the net efficiencies. Moreover, energy uses for CO_2 transport (though relatively minor) and CO_2 injection may contribute substantially to total GHG emission per kWh especially in case of deeper underground storages. This leads to total GHG emission rates which are in the range 2.5 to nearly 4 times higher than those emitted at the PC power plant with capture in case of pre- and post-combustion hard coal technologies, and 2 to 2.5 times higher for pre- and post-combustion lignite PC technologies (in case of lignite the emissions from plant upstream are relatively small due to low CH_4 emissions at mine and because plants are mine-mouth). The case of oxyfuel combustion looks naturally better for the higher capture efficiency, leading to total lower cumulative GHG emission rates. However, due to the very low or nearly zero CO_2 emissions at the power plants with CO_2 capture, the cumulative emissions from the rest of the chain are many times higher.

The graphs on figure 2 show the contribution analysis of stages of the hard coal chain and CO_2 management to cumulative GHG emission per unit of electricity at busbar for the case of PC hard coal with CCS. The legend points out at values for systems with CCS piled up in the given sequence: coal supply, power plant (PP) operation, PP infrastructure (including construction and dismantling), and CO_2 transport and storage (T&S). GHG results are given for the most optimistic scenario for PC hard coal without and with CO_2 capture (post- and oxyfuel-combustion) and minimum (analyzed) CO_2 transport of 200 km and minimum 800 m injection depth aquifer and for the maximum (analyzed) transport distance of 400 km with intermediate pumping and maximum (analyzed) injection depth of 2500 m in a depleted gas field. The influence of post-combustion vs. oxyfuel-combustion technologies (difference in CO_2 removal efficiency [2] and chemicals use for CO_2 separation [5,6]), the coal supply upstream chain, and the T&S in the worst (analyzed) case of 400 km CO_2 transport and 2500 m injection depth in depleted gas field discussed above is clearly illustrated.

Total fuel consumption increases due to higher energy uses at the power plant with CO_2 capture compared to the auxiliary power for a unit w/o CO_2 capture, as illustrated in figure 1. This fact and the energy expenditures for transport (though generally minor) and injection of supercritical CO_2 are directly reflected in the increase of CED fossil for systems with CCS, which ultimately leads to a faster exploitation of non-renewable fossil energy resources. The first applications of CCS in the next ten years will exhibit an increase of CED per kWh at busbar compared to a PC w/o CO_2 capture of 33% to 44% for hard coal, and 35% to 49% for lignite. However, likely improvements in the efficiency of PC and capture technologies may lead to reducing such CED increase compared to corresponding power plants w/o CO_2 capture to 18% to 31% for hard coal, and 16% to 35% for lignite, around year 2050.

Application of EI'99(H,A) [10] shows that CCS technologies may even substantially worsen the (calculated) overall environmental score per unit of electricity output in case of hard coal and lignite post-combustion PC technologies (see figure 1). For hard coal (see two examples in figure 3) this effect is primarily due to the increase of fossil fuel resource consumption (including consumption for electricity uses for CO_2 pumping, modeled using the UCTE mix year 2000 [4]), and the increase of respiratory inorganics because of plant efficiency decrease and material uses for CO_2 separation after post-combustion may lead to reduction of the total score, though in the case of hard coal the latter reduction may not be large. For CSS, also increased are acidification/eutrophication as well as carcinogens midpoint scores (again due to power plant efficiency reduction and material consumption for CO_2 separation), but they remain lower contributors to the total, endpoint EI score. The increases mentioned above may more than compensate the decrease of points for climate change. However, these calculations cannot be conclusive because many uncertainties for the LCA still exist especially on the consumption rate of chemicals (and to a minor extent the uncertainties on the power plant construction materials) and type/manufacturing process of chemicals for post-combustion CO_2 separation technologies (for year 2050 mostly speculative).

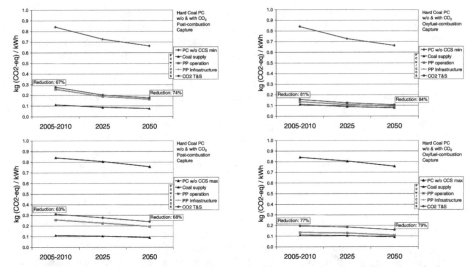

Figure 2. GHG results for PC hard coal without and with post-combustion (left) and oxyfuel-combustion (right) CO_2 capture; most optimistic (VO) scenario (top) with minimum CO_2 transport of 200 km and 800 m injection depth aquifer vs. pessimistic (Pe) scenario (bottom) with maximum CO_2 transport of 400 km and 2500 m injection depth depleted gas field. Contribution analysis for the case with CCS.

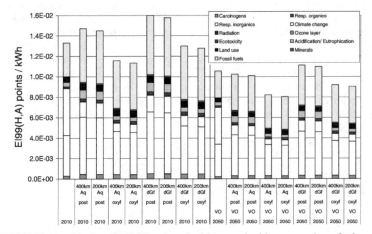

Figure 3. EI'99(H,A) midpoints for PC hard coal without and with post- and oxyfuel-combustion CO_2 capture and two cases for average CO_2 transport and injection depth: min = 200 km and 800 m depth aquifer (Aq); max = 400 km and 2500 m depth depleted gas field (dGf). Only case in year 2010 and most optimistic case (VO) in year 2050 are shown as illustration.

In general, the conclusions of the assessment for coal energy chains with PC and CCS are as follows. Important GHG reductions per unit of electricity supplied to the grid can be obtained for coal systems with CCS but contributions from the hard coal upstream chain increase compared to the PC w/o CCS proportionally to the net efficiency ratios, and extra emissions are generated because of energy and material expenditures for CCS.

Therefore, the GHG retention per unit of electricity is not as dramatic as claimed by the power plant constructors. Higher fossil resources are used for the same electricity output. Overall environmental damage may likely be higher for post-combustion coal technologies than for the system without CCS, unless very high plant efficiencies were obtained and optimal conditions were taken for transport and storage of CO_2. In case of oxyfuel coal combustion, overall environmental damage may likely be comparable to the system without CCS and possibly lower in case of near and not very deep CO_2 storages, more so for the case of lignite PC.

Hard Coal an Lignite IGCC

Figure 4 shows synoptically the results obtained for the three aggregated indicators for the energy chains with IGCC fuelled by hard coal (left) and lignite (right). It should be noted that the efficiency range for Pe-VO has been defined more narrowly than for PC, whereas the efficiency decrease of an IGCC without CCS from today to year 2025 is more dramatic than for PC in the case of pessimistic (Pe) estimations. This assumption on IGCC determines a sharp decrease of all indicators in year 2025 and 2050 for the systems w/o CCS. GHG decrease with CCS by 71% to 75% for hard coal, and 79% to 83% for lignite systems in year 2025-2050 compared to the energy chains with power plants w/o CO_2 capture for the respective year and scenario. As for any fossil technology with CCS, the CED increases with respect to the plant w/o CCS because of the increased auxiliary power necessary for the capture and compression of the CO_2. In general the range of EI'99(H,A) points for the systems with CCS is somewhat lower than for units w/o CCS, but yet for the case of longer transport and deeper storage of CO_2 the points may be very near, i.e. the overall potential damage be comparable for the about compensating effects on different midpoints similarly to what discussed above for the PC cases. Therefore, also for IGCC with CCS similar conclusions as for PC can be derived: large GHG reductions are obtained but not as dramatic as claimed by the power plant constructors; higher fossil resources are used for the same electricity output; overall environmental damage comparable to systems w/o CCS are likely, unless optimal conditions were taken for transport and storage of CO_2.

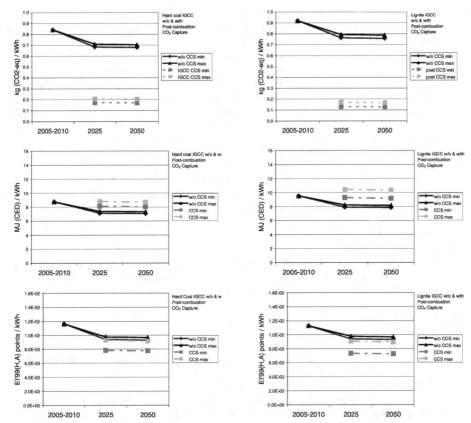

Figure 4. Greenhouse gas emission, Cumulative Energy Demand non-renewable, and EI'99(H,A) points (from top to bottom, respectively) results for IGCC hard coal and lignite (left and right, respectively) without CO_2 capture and with pre-combustion CO_2 capture on each graph) and two cases for assumed European average CO_2 transport and injection depth (min = 200 km and 800 m depth aquifer; max = 400 km and 2500 m depth depleted gas field).

CONCLUSIONS

The analysis of the results obtained applying three aggregation methods (GHG, CED, and EI'99(H,A)) show that although serving CO_2 emission reduction at fossil power plants, application of CCS technology introduces own GHG emissions due to material and energy uses and increases the upstream emissions per unit of electricity delivered to the grid. Moreover, the introduction of CCS may generally not improve the overall environmental efficiency of the systems.

Application of EI'99(H,A) [10] shows that CCS technologies may even worsen the (calculated) overall environmental score per unit of electricity output in case of hard coal and lignite post-combustion technologies for PC.

The increase of total fuel consumption, directly reflected in the increase of CED fossil per unit of net electricity output, leads to a faster exploitation of non-renewable energy resources.

Sensitivity analyses should be performed on the various key parameters, in particular: energy uses in CO_2 transport and storage (i.e. distance from power plant to injection point and depth and overpressure for injection into geological strata); materials (plant and operational) for CO_2 and O_2 (for oxyfuel-combustion) separation; CO_2 reduction potentials in the hard coal and natural gas supply chains.

Estimation of external costs should be performed at the conclusion of the NEEDS project, quantifying the total environmental burdens. It can be anticipated that the total external costs related to pollution are largely depending on the damage factor attributed to GHG, which greatly varies according to different modeling. Depending on this value, the fossil systems may be more or less penalized in comparison to renewable and nuclear energy.

ACKNOWLEDGMENTS

The project NEEDS (New Energy Externalities Developments for Sustainability) is financed under the 6th RTD (Research, Technology Development and Demonstration) Framework Programme of the European Union.

REFERENCES

1. NEEDS Project, European Commission, http://www.needs-project.org/
2. C. Hendriks, "Carbon Capture and Storage", UNFCCC Secretariat Financial and Technical Support Programme, Draft August 23, 2007, unfccc.int/files/cooperation_and_support/financial_mechanism/application/pdf/hendriks.pdf.
3. R. Dones, C. Bauer, T. Heck, O. Mayer-Spohn, and M. Blesl, "Final technical paper on technical data, costs and life cycle inventories of advanced fossil fuels", European Commission, to be issued 2008.
4. www.ecoinvent.ch
5. IPCC, "Carbon Dioxide Capture and Storage" Cambridge University Press, New York (2005).
6. E.S. Rubin, C. Chen, and A.B. Rao, "Cost and performance of fossil fuel power plants with CO2 capture and storage", Energy Policy 35 (2007) 4444–4454.
7. Wildbolz C., "Life Cycle Assessment of Selected Technologies for CO2 Transport and Sequestration", Diploma Thesis No. 2007MS05, Department Bau, Umwelt und Geomatik, Institute of Environmental Engineering (IfU), ETHZ, Zurich (July 2007).
8. Doka G., "Critical Review of 'Life Cycle Assessment of Selected Technologies for CO2 Transport and Sequestration' Diploma Thesis No. 2007MS05 by C. Wildbolz", Zurich (September 2007).
9. Hendriks C., Graus W., van Bergen F., "Global Carbon Dioxide Storage Potential and Costs". ECOFYS and TNO, Netherlands (2004). http://www.ecofys.com/com/publications/documents/GlobalCarbonDioxideStorage.pdf
10. M. Goedkoop, S. Effting and M.Collignon, "The Eco-indicator 99. A damage oriented method for Life Cycle Impact Assessment." PRé Consultants BV, Amersfoort NL (2001).

Mater. Res. Soc. Symp. Proc. Vol. 1041 © 2008 Materials Research Society   1041-R05-02

Integration of Land Use Aspects into Life Cycle Assessment at the Example of Biofuels

Michael Held, and Ulrike Bos
Life Cycle Engineering, Universitaet Stuttgart, Chair of Building Physics (LBP), Hauptstrasse 113, Leinfelden-Echterdingen, 70771, Germany

ABSTRACT

It is well known that the transport sector causes significant environmental impacts worldwide and as a consequence influences the results of Life Cycle Assessment (LCA) studies. Today's fuels are dominated by crude oil derived fuels. In Europe currently 98 % of the road transportation is based on such crude oil derived fuels. Similar ratios can be observed e.g. in the US and other countries. In addition to the environmental impacts, the high dependency on the imports of fossil fuels motivates most European countries to investigate in other than fossil fuel based transport systems. Therefore the European Commission presented an action plan including a strategy with the objective to substitute 20% of crude oil derived fuels by alternative fuel until 2020. To achieve these goals, actions to reduce the import dependency of fuels, the usage of non renewable (fossil) resources and the environmental burdens connected to the use of fuel / propulsion systems have to be addressed. Besides, the energy carrier mix has to be broadened. Especially alternative fuels from renewable resources, BtL (Biomass to Liquid) are supposed to have a high potential.

Recent developments show, that there is a variety of options for fuels available as well as for propulsion technologies that utilize fuels based on renewable resources. It is therefore of key importance to select and promote the fuel/ propulsion system technology which is most beneficiary for a country or region from an environmental but also from an economic and social perspective. For such a sustainability evaluation it is essential to consider the local/regional boundary conditions such as availability of fuel resources, major pollution issues which need to be addressed, supply of secondary energy (e.g. power) etc. LCA is therefore a suitable approach to evaluate and compare different options, due to its transparent consideration of all life cycle stages.

Besides the environmental impacts and resource consumption which are addressed in LCA considerations the needed land is another important aspect when talking about biomass as a resource. As land is a scarce resource that is used for all industry sectors there is a need to address this issue also in LCA. Up to now, no commonly agreed upon methods exist which allow the integration of land use aspects in a consistent way into LCA Software and Database. Currently at LBP-GaBi, University of Stuttgart together with PE International, a method is developed to integrate land use aspects into LCA. Backward processes are now implemented in an applicable way into a LCA database system.

This Paper describes the main approach of the developed methodology for land use consideration within LCA.

INTRODUCTION

Why land use Consideration?

Every day the equivalent area of 125 soccer fields (1 km^2) natural land is sealed in Germany. The national sustainability strategy has the aim to reduce the daily consumption to 0.3 km^2. Not only Germany is affected by a rising use of natural land. Any growing economy has the need of more useable land for housing, transportation, services and industry. That means land is necessary to produce goods and nutrition to guarantee our life on earth as.

Particularly countries with growing population are faced to a fast increasing demand of useable land but this resource is scarce. Therefore the interest of addressing the subject of land use in politics as well as in science is very high.

This is the reason why the incorporation of land use into Life Cycle Assessment (LCA) is an important subject. It should be addressed as an impact category in LCA methodology which enables the evaluation of environmental burdens of products or product systems.

METHOD TO IMPLEMENT LAND USE CONSIDERATION INTO LCA SOFTWARE SYSTEM GABI 4

Through the anthropogenic land use damage to ecological functions can occur and can endanger ecosystems. Therefore the resource land shall be addressed in a sustainability assessment of products, product groups or services. The first challenge is the provision of adequate parameters or indicators to describe land use.

At the University Stuttgart, Department of Life Cycle Engineering, a method to implement land use in the GaBi LCA software system is under development. The method land use is divided into a transformation and occupation phase. For both phases, the following parameters are calculated for land intensive processes: Physical and chemical filtration, mechanical filtration, erosion resistance, groundwater regeneration rate, and net primary biomass production.

Key parameters

The supply with a consistent set of parameters to describe land use is not trivial. There are several demands on these parameters that have to be met:

First of all the availability of data and the geographical resolution of data for these parameters is important. For example, is data available for land use indicators addressing the bio-geographical differentiation? Or can small-scale changes of geographical units be described properly? It is obvious that the more detailed and local data is available the more exact the result of the modeling concerning land use can be. That means also, a pragmatic procedure regarding the relevance of data and parameters has to be found because data research has to be in a suitable and reasonable manner.

Further more there are demands on Life Cycle Inventory (LCI) modeling regarding the qualities of parameters. LCI data for land use needs to be sumable and scalable over the process-chain (like other LCI data as well). The key parameters have to be provided in a consistent way according to elementary flows. Besides, the question has to be answered which processes are relevant at all in LCA regarding land use. Is it necessary to model the land use aspects of any process, whatever it is a mining process or a chemical process?

Lastly demands on the land use indicators themselves are playing a decisive role: How to deal with indirect dependency between various land use key parameters? How to create region specific average values? Are soil quality, biodiversity and biotic production potential addressed accordingly?

All these issues have to be taken into account when creating land use parameters.
A further challenge is the correct description of the different phases of land use. According to the scientific community -dealing with land use in LCA- land use can be classified into two phases, a transformation phase and an occupation phase.

Transformation
The transformation describes the land properties that are modified during the use of the land. For example the change of the use of land from a natural forest to a farm land or open pit mine and after the use phase back into a secondary forest.

Occupation
Occupation describes the maintenance of land properties. For example during the use phase of an open pit mine the land is occupied and the land cannot develop a natural succession. It can be assumed that the status of the land is constant during the occupation phase.

Taking all these aspects into account, for building up a method to model the impact assessment of land use, quantitative information is necessary that describes the soil quality and biotic production potential. Also transformation and occupation effects of land use have to be addressed.
At the University Stuttgart, Department of Life Cycle Engineering, an ecosystem function based classification of anthropogenic land use in process chains (effects of land use are quantified by looking at the effects of certain landscape functions and potentials) is currently developed. A sophisticated tool calculates a set of key parameters for the ecosystem functions which can be entered as inventory data based on [Baitz 2002].

The following ecosystem key parameters are used in the tool to calculate the set of key parameters that are then edited into the GaBi-Software system:

1. **Erosion resistance:** That means the capability of soil to prevent soil loss.
2. **Physical and chemical filtration:** Physical and chemical filtration means the ability of soil to absorb dissolved substances from the soil solution to prevent pollutants entering the soil matrix (characteristic value: cation exchange capacity).
3. **Mechanical filtration:** Under mechanical filtration the mechanical ability of soils to clean a suspension through the binding of pollutants on soil particles is understood.
4. **Ground water regeneration rate**: This is the capacity to regenerate groundwater.
5. **Net primary biomass production**: Net primary biomass production is the ability of the ecosystem to produce biomass.

So far an applicable land use method for generating LCI results is developed, based on [Baitz 2002].
The input background data for the tool is available for many countries.
Also a set of key parameters as inventory data is calculated for several land consuming processes.

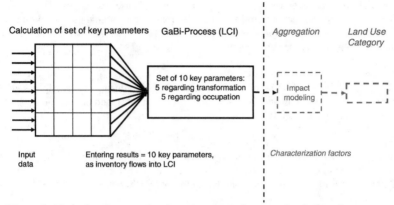

Figure 1: Method to integrate land use inventory data into the GaBi software system

The figure shows the implementation of land use indicators into the GaBi-software system. At the moment a set of 10 key parameters are calculated (until the dashed vertical line). Function-based "weighting" to aggregate the set of key parameters to land use indicators has to be done.

EXAMPLE: LAND USE OF THE CULTIVATION OF ENERGY PLANTS FOR BIOFUEL PRODUCTION

The following example shows the land use inventory data (transformation) for the cultivation of energy plants on agrarian land for two alternative land use types as starting systems: mixed forest and post-mining reclaimed land. The objective of the example is to present the quantitative difference between the transformation of mixed forest to agrarian land and the transformation of a mining area to agrarian land to use this land for Biofuel production.
The consideration is done for the cultivation of 1ha ($10000m^2$) energy plants on respective land types with a cultivation period of 15 years and deciduous forest as reference system that means the land use type without any anthropogenic intervention.
The impacts of the cultivation of energy plants on the land types to transformation indicators are presented in the figures 2-6. Negative values describe an improvement regarding to the starting land use type.
The results show that the cultivation of energy plants on post-mining reclaimed land leads to an improvement for all considered transformation indicators whereas the cultivation on mixed forest leads to degradation in erosion resistance, ground water regeneration rate and net primary biomass production. The reason for this result is the clear improvement of the land: a mining area does not present any working soil functions as the prevention of erosion or the mechanical filtration etc. Therefore, an agrarian land would be an improvement. However the transformation from mixed forest to agrarian land presents in some cases degradation, e. g. there is more erosion and less biomass production as in a forest.

In conclusion this example proves the developed land use methodology. It enables the provision of quantitative inventory data for indicators to be used within LCA databases and software systems and forms the basis for comprehensive land use considerations as a part of LCA.

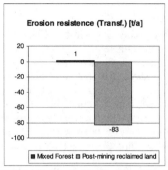

Figure 2: Erosion resistance (transformation)

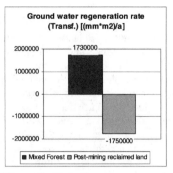

Figure 3: Ground water regeneration rate (transformation)

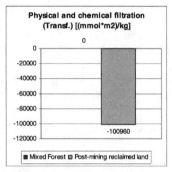

Figure 4: Physical and chemical filtration (transformation)

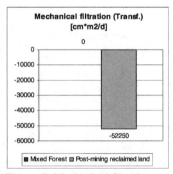

Figure 5: Mechanical filtration (transformation)

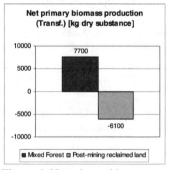

Figure 6: Net primary biomass production (transformation)

163

CONCLUSION

Due to growing relevance land use is a topic that should be considered within Life Cycle Assessment. All products and services comprise land use aspects, therefore the implementation of land use considerations requires a consistent approach and comprehensive methodology, based on quantifiable data that is applicable to all industries. The department Life Cycle Engineering and PE International developed a land use method that allows the quantification and implementation of land use data in a consistent way into the software system GaBi 4. The implementation of land use inventory data into GaBi 4 software system is currently in progress. After setting up a tool to calculate land use inventory data for LCA, the next step has to be the impact modeling and the development of a function-based "weighting" to aggregate the set of key parameters to land use indicator(s).

In this connection it is very important to be consistent to international scientific discussions in the LCA community (e.g. UNEP-SETAC task force), which is already taken into account.

REFERENCES

1. M. Baitz: Die Bedeutung der funktionsbasierten Charakterisierung von Flächen-Inanspruchnahme in industriellen Prozesskettenanalysen, Dissertation, Universität Stuttgart, 2002.

2. M. Baitz; J. Kreissig; M. Wolf: Method for integrating Land Use into Life Cycle Assessment (LCA), Forstwissenschaftliches Centralblatt 119, 2000, Blackwell Wissenschafts-Verlag, 128-149.

3. ISO/FDIS 14040 Environmental Management – Life Cycle Assessment – Principles and Framework, 2006.

4. IKP, PE: GaBi 4 Software-System and Databases for Life Cycle Engineering. Copyright, TM Stuttgart, Echterdingen.

5. Seidel, U.; Braune, A.; Fischer, M.; Baitz, M.; Kreissig, J.: Integrating Land Use into LCA Methodology, SETAC North America Annual Meeting; 5 - 9 November 2006, Montréal, Canada.

6. Schuller O.; Seidel U.; Makishi C.: Addressing Land Use in Life Cycle Assessment - Cilca, Conference 2007 in Sao Paulo, 2007.

7. Seidel, U.; Braune, A.; Fischer, M.; Kreissig, J.; Baitz, M.: Land Use indicators in LCA – demonstrating an applicable approach (poster), SETAC Europe 17th Annual Meeting, Porto, May 20-24, 2007.

8. Bos (Seidel), U.: Implementierung von Flächeninanspruchnahme in die Ökobilanz, Vortrag Ökobilanzwerkstatt Netzwerk Lebenszyklusdaten Bad Urach, 26.-27. September 2007.

Mater. Res. Soc. Symp. Proc. Vol. 1041 © 2008 Materials Research Society   1041-R05-03

The Fuel Cycles of Electricity Generation: A Comparison of Land Use

Hyung Chul Kim[1], and Vasilis Fthenakis[1,2]
[1]PV Environmental Research Center, Brookhaven National Laboratory, Upton, NY, 11973
[2]Center for Life Cycle Analysis, Columbia University, New York, NY, 10027

ABSTRACT

We investigate the area of land used and/or transformed during conventional (i.e., coal, natural gas and nuclear), and renewable fuel cycles (i.e., photovoltaics, wind, biomass, and geothermal). Both direct and indirect land use/transformation are examined in a life cycle framework. For average US insolation, the photovoltaic fuel cycle disturbs the least amount of land per GWh among renewable options, requiring less area than the coal fuel cycle. Renewable technologies could harvest infinite amount of energy per unit area and eliminates the need for restoring disturbed mine lands. Further investigations would be necessary for secondary and accidental land disturbance by conventional fuel cycles through transport of effluents and emissions to adjacent land.

INTRODUCTION

Renewable energy sources such as photovoltaics (PV) and biomass are often criticized as requiring a large amount of land compared to other energy options [1, 2]. However, life-cycle based land assessments present a different picture; conventional fossil fuel cycles require large amounts of land when we accounting the land for fuel extraction [3, 4]. In this article we review the life cycle land use for conventional fuel cycles (i.e. coal, nuclear, and natural gas), and renewable fuel cycles, (i.e. PV, wind, geothermal and biomass). Land use metrics are parsed into land transformation (unit: m^2) and land occupation (unit: m^2 x year). The former, which is employed in this study, indicates the area of land altered from a reference state while the latter indicates the land area and the duration of occupation. Direct land usage was compiled from US Department of Energy (DOE) reports while indirect land usage was compiled from the life cycle assessments and energy data. Land use impacts across electricity generation options were compared.

COAL

For the coal fuel cycle, the direct land transformation is primarily related to the coal extraction, electricity generation, and waste disposal stages while the indirect land use refers to the upstream land use associated with energy and materials inputs during the fuel cycle. The sources of direct land transformation data include Ecoinvent, a commercial database, an LCA literature and DOE reports [3-7]. The land transformation factors in the Ecoinvent database were used for estimating the indirect land usages. Land use statistics during coal mining vary with factors including heating value, seam thickness, and mining methods. Surface mining in the Western US tends to disturb less area (per unit coal mined) than in other areas due to the thick seams, 2-9 m. Central states where seam thickness is only 0.5-0.7 m, transform the largest area for the same amount of coal mined (Figure 1). On the other hand, underground mining transforms land mostly indirect paths. Wood usage for supporting underground coal mines accounts for the majority of indirect land transformation. Currently in the US, about 70% of coal is mined from surface [8]. Table 1 summarizes the conditions for the direct land use calculation of the US mining methods.

For operating a coal power plant, land is required for facilities including powerhouse, switchyard, stacks, precipitators, walkways, coal storage, and cooling towers. The size of a coal power plant highly varies; a typical 1000 MW capacity plant requires between 330-1000 acres [6] which translated into 6-18 m^2/GWh of transformed land based on a capacity factor of 85%. Another study based on a 500-MW power plant located in the Eastern US estimated 32 m^2/GWh of land transformation. On the other hand, coal-fired power plant generates a significant amount of ash and sludge during operation. Disposing the solid wastes account for 2-11 m^2/GWh, 50% each for ash and sludge, in the US condition.

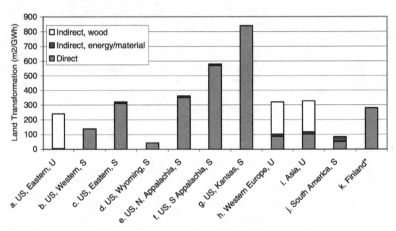

Figure 1. Land transformation in the coal mining stage. U= underground mining, S=surface mining, *mining type and area of indirect land use is unknown. Sources: a-[3, 5]; b-[5]; c-[4]; d-g- [6]; h-j- adapted from [3]; k-[9].

Table 1. Conditions for the US coal mining methods

	Eastern Underground [3, 5]	Western Surface [5]	Eastern Surface [4]
Land transformation (acre), fixed	6	27	30
Land transformation (acre), annual	1.5	535	365
Lifetime of mine (yrs)	23	40	30
Coal mined (million t/yr)	1.4	8.8	2.3
Electricity generated* (TWh/yr)	3.1	15.6	4.8

*Based on electricity conversion efficiency of 0.35; Loss during preparation=0.25

NATURAL GAS

The natural gas fuel cycle consists of extraction and purification, transmission, storage, and electricity generation. The energy and materials usage data are from DOE's description of natural gas fuel cycle [5], which subsequently converted to land transformation equivalence using the Ecoinvent database. The on shore extraction method requires a significant amount of land use both directly and indirectly. The indirect land use is primarily (>90%) related to the diesel fuel usage for drilling 1500 m deep with a 67% of chance of success for each of 120 gas wells, a typical size for a single production field. On the other hand, off shore extraction, which accounts for around 20% of the volume produced during 2003-2005 in the US, requires a negligible amount of land transformation although a large area of water surface (140 m^2/GWh) could be withdrawn from fishing. For the gas transmission stage, it is assumed that 970 km underground pipeline is constructed [5]. The land above the pipeline is designated as "right-of-way" for the gas transmission up to 20 m wide for public safety, prohibiting land development and limiting activities [10]. Note that natural gas is often stored in depleted gas wells that the land transformation for gas storage stage is not added to the total as shown later in this paper. The direct land use for a natural gas power plant is smaller than a coal-fired plant probably because it does not need large fuel storage or emission control structures like precipitator.

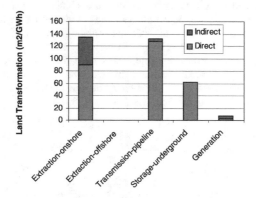

Figure 2. Land transformation during the natural gas fuel cycle [3, 5]

NUCLEAR

We analyzed the direct and indirect land use transformation for the US nuclear fuel cycle. The direct land cover by nuclear power plants accounts for the majority (~60%) of land transformation. A nuclear power plant typically occupies more land than a coal-fired plant (52 m^2/GWh vs. 32 m^2/GWh) due to safety-related exclusion and barrier space. Although a permanent storage for spent nuclear fuel has not been constructed yet, we estimated the land transformed per GWh of electricity generated based on the planning of Yucca mountain repository which will accommodate the spent fuel generated until 2011 in the US. To meet the Nuclear Regulatory Commission licensing requirements, an area of 150,000 acres will be permanently withdrawn isolating the site from public [11]. This would accommodate, by 2011, the total amount of electricity generated by nuclear power plants in the US, that is abound 21,000 TWh [8]. Accordingly, 29 m^2/GWh of land will be needed for the spent fuel disposal.

PHOTOVOLTAICS

The recent progress in PV module and balance of system (BOS) technologies have shown that a large-scale solar electric power plant can provide electric energy for a competitive cost in the near future. The PV fuel cycle consists of materials acquisition, module production, operation and maintenance, and disposal. First, we estimated the direct land use effect for a ground mounted solar PV based on the actual solar panel configuration. The average US insolation of 1800 kWh/m^2/yr is used for this analysis although most commercial power plants are likely to be located in a high insolation area of the South West with insolation on south facing, latitude tilt of about 2300 kWh/m^2 /yr. The lifetime of a PV power plant will be virtually infinite as no major structures or machines, for example, generators, reactors, or cooling systems that need to be demolished for safety or economic reasons. As shown in Figure 3, the land requirement of PV is mostly related to the direct use of land, which in contrast to the land use of fossil fuel cycles, can be used indefinitely. Nonetheless, a lifetime of 60 years was used for this analysis which is the life expectancy for the structural components of BOS in the TEP plant [12]. Figure 4 illustrates the effect of power plant lifetime in measuring the land required, by plotting the land transformed against the lifetime of power plant. A packing factor of 2.5 has been used for the direct land use effect [13]. The direct land transformation will become smaller with a longer lifetime of the operation as the land area is normalized by increasing amount of cumulative electricity generated over years. Indirect land usages are estimated from the life cycle inventories of PV modules and BOS components estimated by recent publications along with the land use factors from the Ecoinvent database [3, 12, 14].

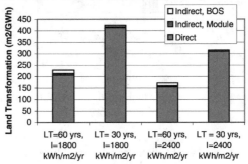

Figure 3. Land transformation for the PV fuel cycle based on ground mounted CdTe PV with an efficiency of 14% and a performance ratio of 0.8. LT = lifetime, I=insolation.

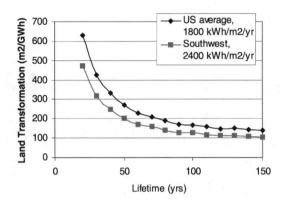

Figure 4. Land transformation for the PV fuel cycle over plant lifetime for CdTe PV with 14% of efficiency.

OTHER RENEWABLES

Electricity generated from biomass is not common yet it is mostly converted to liquid fuel (e.g. ethanol) for combustion. For electricity generation purposes, growing woody biomass (e.g. willow, popular, etc) is currently being explored. The harvesting rate of biomass varies depending on location and species ranging from 1.3 to 4.9 dry tons/acre/yr [2, 15]. An early study estimated a similar range 4900-7900 m^2/GWh depending on the processing option, i.e., gasification, direct fire, and co-firing [13]. A recent study based on willow biomass reports 6100-8100 m^2/GWh of land requirement when normalized for 60 years of lifetime [16]. For a large-scale wind farm, the land requirement varies substantially with the configuration of turbines, ranging 600-1800 m^2/GWh with the same normalization factor [13]. For the geothermal fuel cycle, the land requirement for power plant is insignificant (4 m^2/GWh) compared with the well field underground, ~700 m^2/GWh if accounted for [13].

DISCUSSIONS AND CONCLUSIONS

The total land area transformed during resource extraction, electricity generation and waste disposal are compared across fuel cycles (Figure 5). PV fuel cycles transform land area comparable to the natural gas cycle and less than the coal fuel cycle based on 60 years of plant lifetime. If PV modules are installed in buildings or structures (e.g., shade) only a minimal amount of land will be indirectly disturbed.

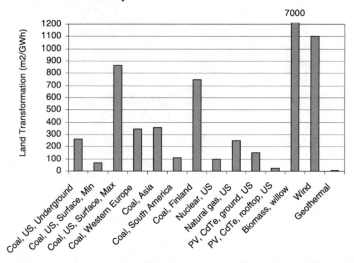

Figure 5. Life cycle land transformation for fuel cycles. The lifetime of PV power plant, wind farm, and crop field is assumed to be 60 years.

Land use of renewable energy sources like PV and biomass poses a different feature from conventional fuel cycles in that they use land in a static way. Once the infrastructure is constructed, renewable fuel cycles will not disturb land afterwards for extracting and transporting resources from surface or underground. This feature eliminates from the renewable fuel cycles the necessity of reclaiming mine lands or securing additional lands for waste disposal. On the other hand, fossil or nuclear fuel cycles need to transform certain amount of land continuously in proportion to the amount of fuel extracted. Restoring land to the original form and productivity takes a long time and often is infeasible. In many cases mines are restored to safe land but transformed to a different contour which is suitable for residential or industrial areas [6]. Our study does not include secondary effects associated with land exploitation that are difficult to quantify. Accounting for secondary effects including water contamination, change of forest ecosystem, and accident-related land contamination would make the PV cycle even better than other fuel cycles. For example, water contamination from coal and uranium mining and from piles of uranium mill tailings would disturb adjacent lands. A recent study shows that Appalachian mountain top mining alters the landscape characteristics of a large area of forest [17]. The boundary between open land and forest disturbs natural ecological goods and services (sunlight, nutrients, moisture, etc.) provided to interior forests, eventually modifying the type of thriving habitats. This illustrates complexity of secondary land use effects and warrants further assessment on the land use of surface coal mining. Additionally, land transformed by accidental conditions especially for the nuclear fuel cycle could change the figures dramatically. The Chernobyl accident contaminated 80 million acres of land with radioactive materials, irreversibly disturbing 1.1 million acres of farmland and forest in Belarus alone [18]. Further investigation would be necessary for these impacts on a regional and global level for a complete picture.

REFERENCES

1. S. Pacala and R. Socolow, Science 305, 968-972 (2004).
2. D. Pimentel, et al., Bioscience 52, 1111-1119 (2002).
3. R. Dones, et al., Sachbilanzen von Energiesystemen. Final report ecoinvent 2000. Volume: 6. Swiss Centre for LCI, PSI, 2003.
4. Meridian Corporation, Energy System Emissions and Material Requirements. Alexandria, VA, 1989.
5. DOE, Energy Technology Characterizations Handbook: Environmental Pollution and Control Factors. U. S. Department of Energy, 1983.
6. K.E. Robeck, et al., Land Use and Energy. Argonne National Laboratory, Argonne, Illinois, 1980. ANL/AA-19.
7. D.V. Spitzley and D.A. Tolle, J. Ind. Ecol. 8, 11-21 (2004).
8. EIA, Annual Energy Review 2006. Energy Information Administration. DOE/EIA-0384(2006).
9. L. Sokka, S. Koskela, and J. Seppälä, Life cycle inventory analysis of hard coal based electricity generation. Finnish Environment Institute, Helsinki, 2005.
10. Pipeline Rights-Of-Way (ROW), website of NW Natural. www.nwnatural.com.
11. Final Environmental Impact Statement for a Geologic Repository for the Disposal of Spent Nuclear Fuel and High-Level Radioactive Waste at Yucca Mountain, Nye County, Nevada. Department of Energy, 2002. DOE/EIS-0250.
12. J.E. Mason, et al., Prog. Photovoltaics. 14, 179-190 (2006).
13. Renewable Energy Technology Characterizations. Department of Energy/EPRI, 1997. TR-109496.
14. E. Alsema and M. de Wild-Scholten. Environmental Impact of Crystalline Silicon Photovoltaic Module Production. in Material Research Society Fall Meeting, Symposium G: Life Cycle Analysis Tools for "Green" Materials and Process Selection. 2005. Boston, MA p. 73-82.
15. P.J. Tharakan, et al., Biomass Bioenerg. 25, 571-580 (2003).
16. D.V. Spitzley and G.A. Keoleian, Life Cycle Environmental and Economic Assessment of Willow Biomass Electricity: A Comparison with Other Renewable and Non-Renewable Sources. Center for Sustainable Systems, University of Michigan, Ann Arbor, MI, 2005. CSS04-05R.
17. J.D. Wickham, et al., Landscape Ecol. 22, 179-187 (2007).
18. R.C. Barbalace, Chernobyl Disaster's Agricultural and Environmental Impact. EnvironmentalChemistry.com, 1999. EnvironmentalChemistry.com/yogi/hazmat/articles/chernobyl2.html.

Mater. Res. Soc. Symp. Proc. Vol. 1041 © 2008 Materials Research Society 1041-R05-05

Wise Energy Investment Decisions--Not Just [kJ out/kJ in]

Lise Laurin
EarthShift, 31 Leach Rd, Kittery, ME, 03904

ABSTRACT

While the best energy solutions may seem obvious to the LCA community, we often see wind turbines voted down for aesthetics and policy makers leaning toward solutions that show poor return, kilojoule per kilojoule. If we are to move forward with wise energy solutions, we will need to broaden our perspective to include the social impacts that influence policy-makers and communities, creating a decision-system that encompasses both social and environmental impacts.

Starting with LCA and Total Cost Assessment, a case study of a biodiesel facility in Vermont begins to incorporate social goals with reduced environmental impacts. We'll then look at other energy systems and how these decision-making tools might be used to bring policy makers, environmentalists, and communities together making wise energy choices for our future.

INTRODUCTION

Politicians and the Life Cycle Assessment (LCA) community are looking at different energy solutions right now for many reasons. These different communities, however, are coming up with different favored solutions. Each group is looking at a different set of criteria for evaluating energy. If the concerns of the LCA community are to be heard, this community needs to first embrace the concerns of the politicians, to create common ground for discussion.

One of the largest drivers for looking at energy solutions is the continually increasing demand for power, with large populations, such as China and India coming on line. Fossil fuel reserves are becoming more and more difficult and costly to extract. For the US and Europe, there is the belief that reliance on off-shore energy sources can lead to national security issues. Increasingly, around the world, global warming concerns are driving energy decisions.

There are a number of alternatives under consideration. For transportation, the top contenders include ethanol, bioethanol, biodiesel, and hydrogen fuel cells. For electricity production there is a wider array of options including, wind, solar PV, clean coal, fuel cells, nuclear, and wood/vegetative scrap. Figure 1 shows a power station which can burn coal, oil, and woody biomass, as an example. Any methodology used to choose between the energy options must be able to incorporate the issues with each as well as handle other options not yet identified.

Figure 1: The author is in a unique position to speak about the social impacts of energy options, with nearly every option less than 15 miles away, or "in her backyard." This is Schiller Power Station in Portsmouth, New Hampshire, which burns oil and coal, but just recently changed one of its boilers over to burn woody biomass.

We can learn something about the limitations of using Life Cycle Assessment by looking at the energy options currently in use in Europe. Figure 2 shows the major impacts of seven different electricity generation methods. We see immediately that hydropower has the least impacts. Yet, in the United States, we are tearing dams down, not building new ones, specifically because of their environmental impacts. The impacts of dam construction on fish migration (or the impacts of wind mills on bird and bat migration) are not taken into account in life cycle assessment because of the site specificity of the impacts.

Comparison of electricity generation options using Impact 2002+

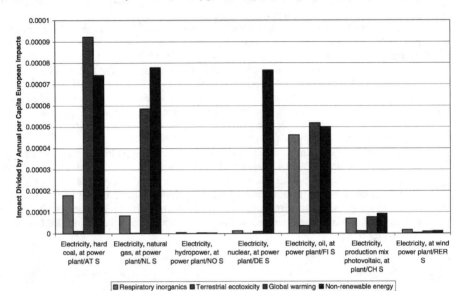

Figure 2: A comparison of the best example of several types of energy production from the ecoinvent 1.3 library [1], using the Impact 2002+ [2] impact assessment method. Data is normalized to annual European per capita impacts. The figure shows only the four categories with the highest normalized impacts.

Looking further, we find that nuclear power is low in most categories, with the exception of non-renewable energy. If we dig into the data for nuclear power, however, we find a small land use (not shown in the figure) that lasts for 80,000 years. This time estimate is important from an overall sustainability aspect. From buildings in Europe, we know that we can build structures to last for 800 or 1000 years. Assuming we have improved our technology since that time, we might estimate that we could build a structure that could protect our descendants from nuclear waste for 8000 years. But 80,000 years? The time span of LCA is too short to take into account the impacts of that nuclear waste after what ever protection it has at that point has been compromised.

While Total Cost Assessment (TCA) doesn't allow us to assess the environmental impacts of these risks, it does allow us to acknowledge that those risks exist, and to give them some value of cost. The combination of LCA and TCA allows the inclusion of many points of view, and provides a framework for assessing both environmental and social impacts of a decision. An example assessment of a biodiesel facility illustrates the power of these techniques to address the needs of community, investors, and the environment.

GOING BEYOND-BIODIESEL IN VERMONT

The people in the state of Vermont, as in many New England communities, are looking to preserve their farmland. To keep these farms in operation, the state needs to add an additional source of revenue. The additional revenue is also important because of the rural nature of the state, with few large industries to support the economy [3]. The state also holds many of the pioneers in the environmental movement and many of its industries have found that reducing impacts is good for business.

Two candidate industries were initially investigated: a cheese factory and a biodiesel plant. The analysis was simplified quickly, however, to include only the biodiesel facility. The analysis did include two different economic models, one with a conventional facility and the other as a cooperative with the farmers. In both cases, corporate goals were the same: sustainability, economic use of raw materials, safe and environmentally sound processes and practices, and, of course, profitability.

The plant design allowed for production of 2.5 million gallons of biodiesel per year, with production to begin 18 months after start of construction. One of the most interesting parts of the analysis followed from the fact that biodiesel can be made from used cooking oil. Figure 3 shows a typical collection tank for used cooking oil, or "yellow grease." Although used cooking oil is always less expensive than virgin oil, as demand for alternative fuels increases, the value and thus the cost also increase. Both uncertain availability and cost volatility needed to be taken into consideration in the model.

Figure 3: A collection tank for used cooking oil outside an Indian take-out restaurant near the author's home feeds local biodiesel production.

In the LCA portion of the model, the analysis was rudimentary and dealt only with virgin oil. Since results with virgin oil showed lower impacts than the petroleum biodiesel it was replacing, there was no need to model the even lower impacts of using used oil.

In the TCA portion of the model, however, the actual availability was varied between 0 and 50% of total feedstock, with a price variance between $0.15 and $0.20 per pound. In addition, although the biodiesel process produces glycerin and animal fodder as co-products, the animal fodder was left out of the system boundaries. The price of biodiesel was modeled as a lognormal distribution with a mean at $1.50 (the prevailing price of diesel at that time) and a standard deviation of $5. (For more details on the models, see reference [4].)

DISCUSSION

Life Cycle Assessment

As our basis for analysis, we looked at two well-known LCA studies of biodiesel: the National Renewable Energy Laboratories (NREL) publication "An Overview of Biodiesel and Petroleum Diesel Life Cycles" for soy-based production of biodiesel and the Berlin-based Institute for Energy and Environmental Research (IFEU) "Life Cycle Assessment of Biodiesel: Update and New Aspects", which we used as an indicator for the canola biodiesel scenario.

The life cycle of biodiesel from virgin canola oil shows a 78% reduction in greenhouse gas production over the life cycle of petroleum-based diesel, or 18 fewer pounds of CO_2 per gallon of fuel consumed. If we look at soy-based biodiesel, it requires slightly more energy to produce than petroleum diesel: 0.23 MJ required /MJ produced for biodiesel, 0.20 MJ required /MJ produced for petroleum diesel. It also leverages fossil fuels, providing 3.2 MJ per MJ of fossil fuel used.

The life cycles of both canola based and soy based biodiesel reduce production of particles, CO, and SOx by reducing the levels at the tailpipe. The absolute reduction amount varies by vehicle, since some vehicles combust at higher temperatures, changing the emissions. NOx is higher for biodiesel, as are total hydrocarbons. The total hydrocarbon emissions, however, are in the field—tailpipe emissions are actually lower; so they may not be as damaging.

Water use is significantly higher for biodiesel, while emissions to water are lower.

The results were essentially positive. The added benefit of using recycled cooking oil will make the project less impacting.

Total Cost Assessment

The Type I & II Costs, or traditional costs, considered in the TCA included the following:
- Plant construction
- Operational costs
- Feedstock costs (including modeling of the less expensive, but price-volatile yellow grease)
- Licensing and Reporting
- Hazardous material handling
- Testing
- Revenues – modeled as negative costs (complex price distribution)

Plant construction costs included safeguards for handling methanol and the hazardous material handling piece included employee training in the handling of methanol to ensure the lowest risk of accidents.

The scenarios included the following:
- Delay due to permitting or other regulatory requirements
- Methanol discharge to air
- Massive Methanol discharge to land
- Employee exposure
- Improper disposal by subcontractor
- Plant Contamination
- Union negotiation
- Product does not meet test criteria

All the scenarios applied to the option where a facility was built. Risks involved with doing nothing, such as the risk of spills of conventional biodiesel, were ignored.

If we look at the net present value (NPV) calculated using the expected values, we see a favorable return on investment. Figure 4 shows the NPV with the value calculated through the year given. With positive results as early as 2007, the return on investment should be rapid.

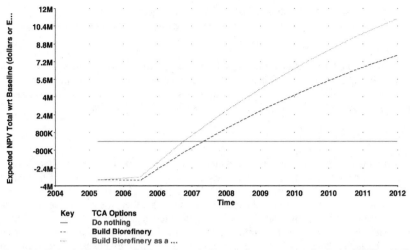

Figure 4: In both biorefinery options (stand alone and as a cooperative), the NPV calculated through 2008 is positive, showing rapid return on investment. Discount rate is set to 0.12.

If we look, however, at all possible scenarios, we find that if things go wrong, the return will not be as high as that shown in Figure 4. Figure 5 shows probable NPV for the non-cooperative biorefinery option based on the simulation. The bottom line represents 5% of the 1000 simulations. Since 95% of the results are better than this line, we see that there is a high probability for success. If we look at the top line, we see that the return has the possibility to be much higher.

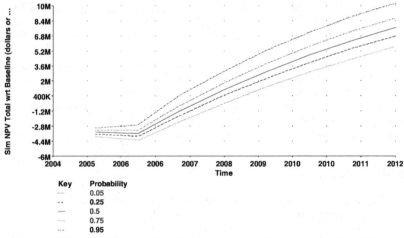

Figure 5: Simulated NPV for the first biorefinery option shows that there is a 95% probability that the NPV will be positive when calculated through about 2009. Discount rate is set to 0.12.

It is interesting that the most variability in these projections comes from variability in feedstock cost and biodiesel revenues. Very little variability is induced by the risk of spills.

The decision

Whether the biodiesel facility was built as a cooperative or not, all the goals of the project were met. The community would maintain its agricultural land, minimize environmental impacts, and add a revenue stream. The corporation would be able to profitably produce a sustainable product using economic raw materials while addressing all safety concerns.

WHAT WAS MISSED?

Looking at concerns of communities and politicians to various forms of alternative energy, there are a number that were not addressed in this particular assessment. Two that apply to biodiesel are the desire of farmers to maintain their quality of life and the issue of food production versus fuel production. Expanding to wind energy, there are still concerns over noise, although current windmills are relatively quiet. Attractiveness is an issue to some. Concern over bird and bat kills is similar to fish migration issues with hydropower. TCA is a powerful tool for this particular issue. If we look at the "do nothing" case, the mercury emissions from traditional electricity generation affects 600,000 in utero children annually [5]. With this many children affected, it is likely that birds and bats are also affected by "doing nothing," perhaps more than by the windmills. Both the risks to children from "doing nothing" and the risks to birds and bats in both cases would be brought out in the analysis.

Solar electricity brings in issues such as development costs and infrastructure costs. In all cases where traditional sources are being replaced, large spills and leaks, which are not considered in current LCA data, should be considered.

Interestingly, the combination of LCA and TCA can handle all of these missing items.

CONCLUSIONS

The combination of LCA and TCA provides a framework for evaluating energy systems that encompasses both environmental and social concerns, allowing parties with focused interests to come together to make the wisest decision. In the case of a biodiesel facility in Vermont, through these methods, the facility proved to be based on sound business practices while meeting the goals of the community. In the broader case of energy systems, the bringing together of disparate parties is critical if the world is to address the concerns before it and build a sustainable future.

ACKNOWLEDGMENTS

Funding for the biodiesel case study was provided by the AIChE Institute for Sustainability through Vermont's Alternative Energy Corporation.

REFERENCES

1. Frischknecht, Rolf, Jungbluth, Niels (2004) "Ecoinvent Overview and Methodology" Published by the Swiss Centre for Life Cycle Inventories, Dübendorf, Switzerland, 2004. http://www.ecoinvent.org/fileadmin/documents/en/01_OverviewAndMethodology.pdf

2. Jolliet O, Margni M, Charles R, Humbert S, Payet J, Rebitzer G, Rosenbaum R (2003): IMPACT 2002+: A New Life Cycle Impact Assessment Methodology. Int J LCA 8 (6) 324 – 330

3. Mulder, Kenneth, Matthews, Allen, "An Ecological Economic Assessment of a Proposed Biodiesel Industry for the State of Vermont", Report Summary for USDA grant NRCS 68-3A75-3-143, Aim 3, December 2004, http://www.uvm.edu/%7Esusagctr/BRDI%20summary.pdf.

4. Laurin, Lise, Norris, Gregory A., Trupia, Sabrina, "Biodiesel in Vermont—the environmental impact and the total cost", Case Study produced for AIChE, July 2005. http://www.earthshift.com/Biorefinery%20Case%20Study.pdf

5. Altschiller, Howard, "N.H. musn't wait for EPA to take action on mercury." Portsmouth Herald, Feb. 16, 2005. http://archive.seacoastonline.com/2005news/02162005/editoria/64817.htm

Note: All web links are current as of January 24, 2008.

Mater. Res. Soc. Symp. Proc. Vol. 1041 © 2008 Materials Research Society   1041-R05-06

Standing the Test of Time: Signals and Noise From Environmental Assessments of Energy Technologies

Björn A. Sandén

Energy and Environment, Environmental Systems Analysis, Chalmers University of Technology, Sven Hultins gata 6, Göteborg, SE-412 96, Sweden

ABSTRACT

The point of view taken here is that systems analysis is a kind of learning process, not data gathering, not decision making, but the production and effective communication of arguments relevant in a particular context. This idea, that the intended application of the result of an assessment has consequences for methodological choices, is beginning to spread in the LCA research community. One problem is that standard LCA methodology is developed to answer questions about environmental impacts of the current production and use of one unit of a product or minor product or process changes. When this methodology, unchanged, is used to provide answers to questions about strategic technology choice, i.e. not decisions that aim at improving a process within an existing technological environment, but with the long-term goal of changing large-scale technological systems, the result could be of little value or misleading. In many cases, LCAs produce more noise than knowledge. This observation seems to be of particular importance for LCAs of energy technologies and for how energy use is treated in all kinds of LCAs. Here, it is suggested that a better understanding of some critical methodological issues related to time, universality, cause-effect relationships, technical maturity and system innovation, could result in better studies that reveal fundamental environmental issues related to the objects of study and reduce the noise from irrelevant information. Examples are given from the technology fields of solar cells, fuel cells, batteries, renewable transport fuels and carbon nanoparticles.

INTRODUCTION

In front of me on my desk there is a graph with many bars that is supposed to say something about the environmental load of a technology. It is marvellously detailed and at a quick look it seems to be based on thorough work. Looking closer the results are strange and I realise that the details are at one level correct but the core of the study is flawed. Too little thought went into the relationship between the question the study was supposed to answer and the design of the study, and too much work went into following the LCA protocol to the bitter end, hammering out details that no one except the author will ever care about, or worse, will be taken as a proxy for a broad technology field and basis for strategic decisions. When I look around, I see LCAs of solar cells that have the same LCA profile as a mix of coal and gas, the very technologies solar cells are to replace. More generally, a multitude of studies seems to do little but reflect more or less arbitrary electricity background systems. I see LCAs of biofuels and new propulsion systems with results that differ by orders of magnitude from study to study. I see bold claims made in political and industrial arenas based on various studies with narrow scopes that would not allow for that kind of generalisations. Within the LCA community (and now also in wider circles) I see endless debates whether it is more "correct" to use marginal data or average data. I see decimals used to signal a precision of atomic clocks in studies based on input variables that are situation dependent, arbitrary and uncertain at the order of magnitude level.

Many studies are not made to stand the test of time, but are made for a small group of people to learn something or to be the basis for a minor decision with short-term implications. That is fine. But in these days when mitigation of climate change is in focus, there is a constant temptation to jump to conclusions about long term and society wide implications. In this paper I will try to make a number of distinctions that I think are relevant for all kinds of LCAs and probably also for other types of environmental assessments. However, I think they are particularly relevant for assessments of energy technologies that want to say something about desirable directions of change. Some ideas are new, but at large the paper is based on a tradition that has evolved at the Division of Environmental Systems Analysis at Chalmers over the last fifteen years, and more specifically on a number of case studies dealing with environmental assessments of different emerging energy technologies that we have conducted over the last ten years. One way to capture the essence is: "it is better to be roughly right than precisely wrong".

TWO POLES OF SYSTEMS ANALYSIS

LCA is a systems analysis methodology. Systems analyses are always caught up in a force field between data and decision. A systems analysis is never a general collection of data that can be used for any decision. Believing that is what we may call a "virgin-scientist" mistake. The relevance of data selection and system boundaries is always dependent on goal and scope, on decision context. On the other hand, a systems analysis is never a complete decision support and it is definitely not the decision itself. There are always things left out and conflicting goals that cannot be resolved by analysis but have to be left for the decision making process. Having the ambition to provide a complete decision support could be called a "super-politician" mistake. Instead, systems analysis (and LCA) is a selection and structuring of data to produce relevant arguments in a specific (decision or learning) context.

There is a resemblance to Heisenberg's uncertainty principle in quantum physics. In Heisenberg's uncertainty principle it is the position and momentum of a particle that cannot be decided precisely at the same time. If you know the location well, you know little about the particle, if you know the speed and direction of the particle you don't know where it is. In systems analysis, a different trade-off is present. If you try to stay very close to data and never reach the level of an argument, the study is of little use for decision making but as you get closer and closer to a full decision support, the scientific basis becomes weaker and weaker. There are many credible positions between the two extremes. There are many ways to select and structure data, but the choices made are neither universally correct, nor arbitrary, but always more or less relevant in a given context, for a given goal and scope [1].

TWO ASPECTS OF TIME

An important distinction that has created some confusion in the LCA community is the one between what is now known as attributional and consequential LCA. For the everyday practitioner it has often boiled down to a question of using marginal or average data. Some years ago the consequential perspective was associated with future oriented studies and the attributional with more accounting retrospective type of studies. Over the last four years we and others have pointed out that the attributional as well as the consequential perspective can be used for various points on a time axis, in history or in the future [2-5].

A consequential perspective investigates the consequences of an intervention. It can be an evaluation of an historical intervention (what were the effects of doing project X?) or an assessment of an intervention that could be made sometime in the future (tomorrow or in thirty years). The studies are retrospective or prospective from the position of the analyst. The object of study (the intervention) is the starting point of a cause-effect chain stretching into the future from the time of the intervention (even if it is in the past from the point of the analyst).

In attributional studies the object of study is instead the end point of a cause-effect chain. (This way of describing attributional LCA differs from references [2-5] but is in fact more general and goes back to a comment made by Ekvall and Tillman [6].) What caused this product or service (what environmental impact made product X possible)? This question applies not to assessments of projects (interventions) but to assessments of products, services and technologies. Also in this case the object of study can be located at different points on a time axis. (What made X possible ten years ago, today or in thirty years?)

There are thus two separate aspects of time: if we from the position of the analyst look back or into the future (retrospective/prospective) and if we from the position of the object of study look back at causes or into the future at effects (attributional/consequential). The relevant choice of position and view differs depending on what is to be investigated. It depends on the question asked. It is also clear that both consequential and attributional assessments can inform decision-making, but that consequential assessment is one step closer to the decision-making pole increasing the burden on the analyst while limiting the role of the decision maker.

TWO IMPLICATIONS OF TECHNOLOGICAL IMMATURITY AND THE ROLE OF STOCKS

The cause-effect chains in attributional and consequential assessments in principle stretch *ad infinitum* into the past and the future, respectively. To be practical, there is a need for cut-off rules. In attributional studies, the production of (at least) two kinds off stocks is normally omitted: physical capital (factories, production equipment) and knowledge. For mature products and technologies, this is not a problem. The flows normally cause larger environmental effects than the production of stocks if averaged out on many products. For mature products, the general definition of attributional as an end-point of cause-effect chains can thus be reduced to a description of the required flows in a given state or temporal window (as done in ref [2-4]). However, for new and emerging products and technologies with small annual and cumulative production volumes, the relative environmental load from building up these stocks could be much larger than the load from flows required for the production of one functional unit. This sometimes causes confusion. My view on this is that in most cases it is more relevant to analyse a (future) situation when the product or technology has reached maturity. However, if the technology's potential is small in relation to required capital and knowledge, the contribution from these stocks could still be significant and relevant to assess.

In consequential studies, changes of stocks are also normally omitted. In a recent paper we differentiate between first, second and third order effects of an intervention [5]. First order effects are direct physical consequences of an intervention. Second order effects are consequences mediated by price signals (supply and demand relationships) on mature markets. The consequences should in principle be calculated as the difference between two equilibrium states with marginal adjustment of supply and demand on different markets [7]. In practice, however, analysts tend to pick a marginal supply technology, for example the marginal electricity production and leave out other marginal adjustments such as reduced electricity consumption in other sectors (for other examples see ref [5]). The idea of a shifted equilibrium takes care of the problem with an infinite cause-effect chain. In the equilibrium economy the main mechanism is negative feedback and the infinite chain converges towards a limited total effect. However, in some cases intervention creates dynamics governed by positive feedback. A small decision can create a snowball (or even a butterfly) effect with large long-term implications. Once again, this is particularly important for immature products and technologies where investments are used to build up stocks of capital and knowledge, which enable economies of scale and learning, decrease costs and expand the use of the product or technology. The environmental implications could thus go far beyond short-term marginal adjustments. For example, we have estimated that this third order effect could be several orders of magnitude larger than the first and second order effects for investments in fuel cell and solar cell systems [5, 8]. We concluded that consequential assessments should not only include effects resulting from marginal change of the current system but also marginal contributions to radical system change.

Taking the step from attributional assessment of products, technologies and services to investigating various consequences of an intervention is an interesting and illustrative exercise but at the same time requires more controversial and uncertain assumptions. In many cases it could be justified to be content with an attributional assessment of a technology in a future state.

TWO TIMES TWO TYPES OF UNIVERSALITY

When an assessment is made of a technology instead of a product from a specific plant, e.g. to support strategic decisions on technology choice, the issue of universality is crucial. Is the case study as presented representative for a broadly defined technology under a wide range of conditions or does the case only represent a narrowly defined product in a specific setting. We can differentiate between two types of universality. The first has to do with the representation of the technology in focus of the study, i.e. the foreground system. The second relates to what context (state, situation) the foreground system is placed in, i.e. the representation of background systems. Within these groups we may differentiate between static and dynamic variation.

Static foreground variation: Within any technology grouping at a certain point in time there is a variation of performance. More generally one may ask how large group of technologies a study represents. Care should be taken not to misrepresent a technology or make conclusions about a broader set of technologies than what is actually assessed.

Dynamic foreground variation: As mentioned in the previous section, the most relevant state to analyse is not always the current state but sometimes also future states. Over time, the technology is developed and too pessimistic or too optimistic assumptions on future technology performance need to be avoided. For example, the performance of a new generation of lithium-ion batteries appears to be superior to the current generation. This will affect any assessment of electric vehicles. Moreover, not only technical development per se, but also the scale of production could change the relevant representation of the foreground system. In an LCA of laboratory scale production of carbon nanotubes, Isaacs et al. estimate the energy requirement to be in the order of 1 TJ/kg [9]. This can be compared to our results that end up in the order of 1 GJ/kg for a future industrial scale production [10].

Static background variation: Few technologies are used in one place only. To assess the generic environmental properties of a technology, the impact of background systems such as electricity production should be clearly shown by sensitivity analysis or 'stylised states'. In a study of wheat ethanol we demonstrated that the emissions of greenhouse gases were six times higher for energy background systems based on coal compared to when they were based on biomass [3, 11].

Dynamic background variation: The same argument is also valid for variations over time. In future states the background systems will be different, but we don't now how. The ethanol example above is valid for the current small market penetration of wheat ethanol in Europe. It is assumed that the main byproduct is used for production of animal fodder. For higher market penetration this market will be saturated and the byproduct could instead be used for heat production. This will dramatically change the environmental load of the ethanol. Thus, for major technological systems, changes of scale may influence background systems. Using the byproduct as heat in the foreground system can also be seen as a way to make the foreground system less dependent of the background system in the analysis. This strategy can be further developed for energy technologies. In the case of biofuels, the produced biofuels itself can replace petrol and diesel as model input for required transport. For solar cells, part of the produced electricity could be used as electricity input and so on. We call this a net-output approach. The good thing is that the inherent properties of the technology are highlighted while the dependence on situation dependent background systems is decreased.

My general recommendation for technology assessments is to look for the inherent properties of the technology, take into account static and dynamic variation of technology foreground parameters and analyse the performance under a range of different background conditions using wide ranges and stylised states instead of one "best" estimate and to critically discuss the relevance of different states (time, scale), where relevance depends on the goal and scope of the study.

TWO ASPECTS OF A PRODUCT

When we move from assessments of specified products to assessments of technologies and long term issues, uncertainty about the relevant definition of the object of study creeps in. The product is an agglomeration of technology and service, of mechanisms and function. When the relevant state for the analysis is located some way into the future this agglomeration does not have to hold any longer. Seen from the service (or functional) perspective, the flexibility of designing new technological systems increases over time. Not only substitution of components is possible but also replacement of complex systems, i.e. the service can be provided in very different ways. Seen from the technology side, technologies can come to be used in novel ways. This has a number of implications: If means to fulfil a certain service (function) is to be assessed, novel alternatives might be more relevant to consider than those currently available. The functional unit should then perhaps be less excluding. For example, biofuels is perhaps a too narrow definition of renewable fuels. Solar electricity combined with electric vehicles could for example reduce in the order of hundred times more CO_2 per acre than any biofuel and use less valuable land. On the other hand, if the environmental profile of a technology is to be assessed, it could be relevant to consider a range of different applications. The same example applies: solar electricity is not only applicable for current electricity markets but also for transportation. For assessments of new materials, such as carbon nanoparticles, the problem of selecting possible and relevant application areas is even harder.

We might have to think in even broader terms. Over longer time spans, new clusters of interrelated technologies tend to emerge. Multiple functions and multiple technologies are combined in new ways. (The case of co-production of ethanol and animal fodder is a minor example this.) If we are going to speak about impact on climate change and deal with timescales of many decades this needs to be considered. Imagine making assessments of the environmental impact of steam engines in the 18th century. In combination with iron, coal, railways, steamships and new ways of organising society, it changed the world, and of course the environment. So did the internal combustion engine and the petrol and diesel oil fractions of petroleum. Today, we could for example ask us what a technology cluster around information technologies, nanotechnology, distributed electricity production and storage, electric vehicles and new spatial and temporal organisation of work and leisure could imply, or a cluster around biotechnology, biofuels, biomaterials, food and food habits. If not issues directly suitable for life cycle assessment, such questions could be relevant for the selection and design of case studies, for the framing of LCAs and for spurring a critical discussion of what kind of conclusions that can be made from LCAs of energy technologies.

CONCLUSION: SIGNALS AND NOISE

This text is not a plea for that only broad forward looking technological assessments are worthwhile. Adding a small piece of data on some specific system is also credible. However, the central question of case studies will never go away: "What is the case study a case of?" When claims are made about environmental pros and cons of technologies, which are to inform decision makers about strategic choices and directions of future development, one has to be careful. There is a risk that one will do nothing but pollute the information space with noise, or worse, with false signals.

What shall one do when the present is irrelevant and the future is uncertain? To stand the test of time, look for the inherent problems and benefits, not the time and site specific. Go for magnitudes rather than decimals. Search for key issues and potential showstoppers. Finally, no matter however careful you are, there is always a risk for misinterpretation and misuse of a study. To do something about that, one needs to get involved in the larger social learning process. Systems analysis is in the end a kind of collective learning process, not data gathering, not decision making, but the production and effective communication of relevant arguments.

ACKNOWLEDGMENTS

Financial support from MISTRA - the Swedish Foundation for Strategic Environmental Research and The Volvo Research and Educational foundations is gratefully acknowledged. I thank Karl Hillman, Magnus Karlström, Anne-Marie Tillman, Duncan Kushnir and Tomas Ekvall for crucial empirical and theoretical input.

REFERENCES

1. A.-M. Tillman, Environmental Impact Assessment Review 20, 113 (2000).
2. M. A. Curran, M. Mann, and G. Norris, Journal of Cleaner Production 13, 853 (2005)
3. K. M. Jonasson and B. A. Sandén, CPM Report No. 2004-6, Chalmers University of Technology, Göteborg, Sweden, 2004.
4. B. A. Sandén, K. M. Jonasson, M. Karlström and A.-M. Tillman., presented at the LCM2005 (Innovation by Life Cycle Management, Barcelona, 2005, Vol 1) pp. 37-41.
5. B. A. Sandén and M. Karlström, Journal of Cleaner Production 15, 1469 (2007).

6. T. Ekvall and A.-M. Tillman, International Journal of Life Cycle Assessment **2**, 155 (1997).
7. T. Ekvall and B. P. Weidema, The International Journal of Life Cycle Assessment **9**, 161 (2004).
8. B. A. Sandén, presented at Workshop on life cycle analysis and recycling of solar modules - the waste challenge, Brussels (European Commission, DG Joint Research Centre, 2004) pp. 31-45.
9. J. A. Isaacs, A. Tanwani, and M. L. Healy, IEEE International Symposium on Electronics and the Environment, 2006, pp. 38-41.
10. D. Kushnir and B. A. Sandén, Journal of Industrial Ecology (submitted for publication) (2007).
11. K. M. Hillman and B. A. Sandén, International Journal of Alternative Propulsion (accepted for publication) (2007).

AUTHOR INDEX

Adachi, Tadaharu, 125
Ahn, Channing C., 75
Alsema, Erik, 3
Araki, Wakako, 125

Barker, Sarah A., 63
Bauer, Christian, 147
Benham, Michael, 63
Berube, Vincent, 51
Blesl, Markus, 147
Bos, Ulrike, 159
Brown, Craig M., 75
Buckley, S. Philip, 63
Burress, Jacob W., 63

Cepel, Raina J., 63
Chen, Xi, 95
Curtis, Calvin, 107

Davis, Mark, 107
de Wild-Scholten, Mariska, 3
Dillon, Anne C., 63, 107
Dones, Roberto, 33, 147
Dresselhaus, Mildred, 51

Engtrakul, Chaiwat, 107

Frankl, Paolo, 13
Frischknecht, Rolf, 33
Fthenakis, Vasilis, 25, 165

Gordon, Michael J., 63
Graetz, Jason, 85
Gualtero, S., 25

Heck, Thomas, 147
Held, Michael, 159

Ilavsky, Jan, 63
Itoh, Kohei, 131

Jones, Kim, 107
Jungbluth, Niels, 33

Kabbour, Houria, 75
Karoglou, M., 43
Kim, Daesuk, 139
Kim, Hyung Chul, 25, 165
Kim, Kiyoung, 139
Kim, Seon Hye, 131
Kim, Yong-Hyun, 107

Lapilli, Cintia M., 63
Laurin, Lise, 173
Licht, Stuart, 119
Liu, Yun, 75

Mayer-Spohn, Oliver, 147
Moon, Byungmoon, 139
Moropoulou, A., 43

Neumann, Dan A., 75

Ohshima, Toshihiro, 131
O'Neill, Kevin, 107

Palyvos, J.A., 43
Panagopoulos, V., 43
Parilla, Philip A., 63, 107
Pfeifer, Peter, 63
Pobst, Jeffrey S., 63

Qiao, Yu, 95

Radke, Darren J., 63
Raugei, Marco, 13
Reilly, James J., 85
Roth, Michael W., 63

Sandén, Björn A., 183
Sasaki, Kazunari, 131
Shah, Parag S., 63
Shin, Jesik, 139
Shiratori, Yusuke, 131
Simpson, Lin, 107
Suppes, Galen J., 63

van der Meulen, R., 25

Wegrzyn, James, 85
Wexler, Carlos, 63
Whitney, Erin, 107
Wood, Mikael B., 63

Yu, Xingwen, 119

Zhang, Shengbai, 107
Zhao, Yufeng, 107

SUBJECT INDEX

adsorption, 63, 75, 107

biomaterial, 173

C, 63

defects, 43

elastic properties, 125
electrical properties, 125
energetic material, 43, 119
energy-storage, 51, 85, 95, 119,
    131, 183
environmentally
    benign, 3, 13, 25, 33, 43, 119,
       147, 165, 173
    protective, 33

Fe, 107

government policy and funding,
    147

H, 51, 85
hydrogenation, 85

nanoscale, 95
nanostructure, 25, 51, 95, 107, 183
neutron scattering, 75

particulate, 173
photovoltaic, 3, 13, 25, 33, 139,
    165, 183
powder processing, 139

Si, 3, 139
storage, 63, 75, 147
surface reaction, 131

thin film, 13

waste management, 165

x-ray photoelectron spectroscopy
    (XPS), 131

Zr, 125